Manual of Leaf Architecture

BETH ELLIS

DOUGLAS C. DALY

LEO J. HICKEY

KIRK R. JOHNSON

JOHN D. MITCHELL

PETER WILF

SCOTT L. WING

Published in Association with The New York Botanical Garden

 THE NEW YORK BOTANICAL GARDEN PRESS

I.C.C. LIBRARY

Comstock Publishing Associates
a division of
Cornell University Press

Ithaca, New York

QK
649
.M326
2009

Copyright © 2009 by Cornell University

All rights reserved. Except for brief quotations in a review, this book, or parts thereof, must not be reproduced in any form without permission in writing from the publisher. For information, address Cornell University Press, Sage House, 512 East State Street, Ithaca, New York 14850.

First published 2009 by Cornell University Press
First printing, Cornell Paperbacks, 2009

Printed in the United States of America

Library of Congress Cataloging-in-Publication Data

Manual of leaf architecture / Beth Ellis ... [et al.].
        p. cm.
 Includes bibliographical references and index.
 ISBN 978-0-8014-7518-4 (pbk. : alk. paper)
 1. Leaves--Morphology--Handbooks, manuals, etc. 2.
Leaves--Anatomy--Handbooks, manuals, etc. I. Ellis, Beth. II.
Title.

 QK649.M326 2009
 575.5'733--dc22

 2008044216

Cornell University Press strives to use environmentally responsible suppliers and materials to the fullest extent possible in the publishing of its books. Such materials include vegetable-based, low-VOC inks and acid-free papers that are recycled, totally chlorine-free, or partly composed of nonwood fibers. For further information, visit our website at www.cornellpress.cornell.edu.

Paperback printing   10 9 8 7 6 5 4 3 2 1

Designed by Margaret McCullough
www.corvusdesignstudio.com

1/10 B&T 29.95

Contents

About the Authors

Beth Ellis is a Research Scientist at the Denver Museum of Nature & Science.

Douglas C. Daly is Director of the Institute of Systematic Botany
at The New York Botanical Garden.

Leo J. Hickey is Professor of Geology at Yale University and Curator of Paleobotany
at Yale Peabody Museum of Natural History.

Kirk R. Johnson is Vice President of Research and Collections and Chief Curator
at the Denver Museum of Nature & Science.

John D. Mitchell is a Research Fellow at The New York Botanical Garden.

Peter Wilf is Associate Professor of Geosciences at Pennsylvania State University.

Scott L. Wing is Research Scientist and Curator in the Department of Paleobiology
at the Smithsonian Institution.

Acknowledgments

KRJ and BE acknowledge support from the National Science Foundation under Grant no. 0345910. LJH acknowledges support from the National Science Foundation under Grant no. 0431258 and from Yale Peabody Museum. PW acknowledges support from the David and Lucile Packard Foundation. Insightful reviews were provided by Robyn Burnham and Lawren Sack. The authors thank the following individuals for their support in producing this document: Bobbi Angell, Amanda Ash, Richard Barclay, Ellen Currano, Regan Dunn, Richard Ellis, Carolina Gómez-Navarro, Katherine Kenyon Henderson, Fabiany Herrera, Rebecca Horwitt, Carol Hutton, Ramesh Laungani, Stefan Little, Mandela Lyon, Dane Miller, Ian Miller, Amy Morey, Daniel Peppe, Sandra Preston, Mary Ellen Roberts, Dana Royer, and Caroline Strömberg.

Illustration credits

The following figures were drawn by Rebecca Horwitt: 1, 8, 10, 14, 22, 23, 24, 25, 26, 27, 28, 29, 30, 32, 33, 34, 35, 36, 37, 38, 39, 40, 42, 43, 44, 45, 46, 47, 48, 49, 50, 84, 85, 86, 87, 125, 126, and 128.

The following figures were drawn by Amanda Ash: 2, 17, 92, 244, 297, 298, 299, 300, and 301.

The following figures were drawn by Bobbi Angell: 13, 124, and 127.

The following figures were provided courtesy of The New York Botanical Garden: 15, 16, 131, 153, 193, 200, 208, 217, 227, 228, 230, 233, 250, 251, 253, 254, 276, 292, 303, 304, Appendix 6, Appendix 16, and Appendix 17.

Figure 51 photograph courtesy of Dennis Stevenson.

Figure 81 photograph courtesy of Dana Royer.

Appendix 18 leaf photograph courtesy of the Denver Museum of Nature & Science.

Other images are included by courtesy of Smithsonian Institution and Yale Peabody Museum of Natural History, being either photographed from the Smithsonian Institution collections that are currently housed at both institutions, or reprinted from the 1999 edition of the *Manual of Leaf Architecture*.

Introduction

Since the time of Linnaeus, comparative analysis of reproductive characters has been the principal morphological technique for identifying and classifying angiosperms (e.g., Takhtajan, 1980; Cronquist, 1981). The Linnaean system and its descendants have been very successful, but there are compelling reasons to increase the use of foliar characters in angiosperm identification and systematics. For example, living tropical plants may flower and fruit infrequently, and reproductive organs may occur only high in the canopy when they are present, making foliar characters more practical for field identification (Gentry, 1993). Even when reproductive organs are available, foliar features can provide information that enhances systematic analyses (Levin, 1986; Keating and Randrianasolo, 1988; Högermann, 1990; Todzia and Keating, 1991; Seetharam and Kotresha, 1998; Roth, 1999; González et al., 2004; Martínez-Millán and Cevallos-Ferriz, 2005; Wilde et al., 2005; Gutiérrez and Katinas, 2006; Doyle, 2007; Manos et al., 2007).

One of the most critical uses of foliar characters is in interpreting the angiosperm fossil record. Although fossil reproductive structures comprise an important source of data (e.g., Friis and Skarby, 1982; Basinger and Dilcher, 1984; Herendeen et al., 1999; Crepet et al., 2004; Friis et al., 2006), compressions and impressions of leaves are the most common macroscopic angiosperm fossils. Because of their abundance, fossil leaves provide a great deal of information about the composition, diversity, and paleoecology of past floras (Chaney and Sanborn, 1933; MacGinitie, 1953; Burnham, 1994; Johnson and Ellis, 2002; Wang and Dilcher, 2006). Furthermore, fossil leaf morphology is widely used to produce estimates of paleoclimatic and paleoenvironmental conditions (Bailey and Sinnott, 1915, 1916; Chaney and Sanborn, 1933; Wolfe, 1971, 1995; Utescher et al., 2000; Jacobs and Herendeen, 2004). Fossil identifications, including those based on leaves, are also used to estimate divergence times of clades (e.g., Richardson et al., 2000, 2001; Renner, 2004; Davis et al., 2005; Uhl et al., 2007).

Working with isolated fossil angiosperm leaves is a long-standing challenge in paleobotany. Late-nineteenth- and early-twentieth-century paleobotanists left a legacy of poorly defined taxa. Most early workers had neither an accepted lexicon for describing leaf form nor knowledge of how leaf features are distributed among living angiosperms (see discussions in Dilcher, 1973; Hill, 1982, 1988). They focused mostly on shape, size, and generalized vein characters that failed to discriminate species or even higher taxa accurately and routinely applied names of living genera to fossils from unrelated fossil genera based on poorly preserved leaves without diagnostic characters. Thus, modern workers inherited a host of misidentified fossil species incorrectly described as *Ficus*, *Populus*, *Aralia*, and other modern genera.

Two recently developed approaches address some of these problems. One method is the

General Leaf Definitions

This section describes the shape, size, surface, organization, and other general features of leaves. Some suites of characters are treated only briefly or are omitted entirely because they have been well described by other researchers. For descriptions of modern leaf surfaces including cuticular morphology, see Dilcher, 1974; and Wilkinson, 1979, pp. 97–117. For more detailed treatment of stipules, stipels, pseudostipules, and phyllotaxy (leaf arrangement), see Bell, 2008; and Keller, 2004. For more detailed treatments of leaf domatia, see Wilkinson, 1979, pp. 132–140; and O'Dowd and Wilson, 1991.

General Orientation Terms

abaxial
Pertaining to the surface of the leaf facing away from the axis of the plant, generally the underside of the leaf (Fig. 1).

adaxial
Pertaining to the surface of the leaf facing toward the axis of the plant, generally the upper surface of the leaf (Fig. 1).

admedial
Toward the midvein (Fig. 2).

apical, distal
Toward the apex (tip) of the leaf (Fig. 2).

basal, proximal
Toward the base of the leaf (Fig. 2).

exmedial
Away from the midvein (Fig. 2).

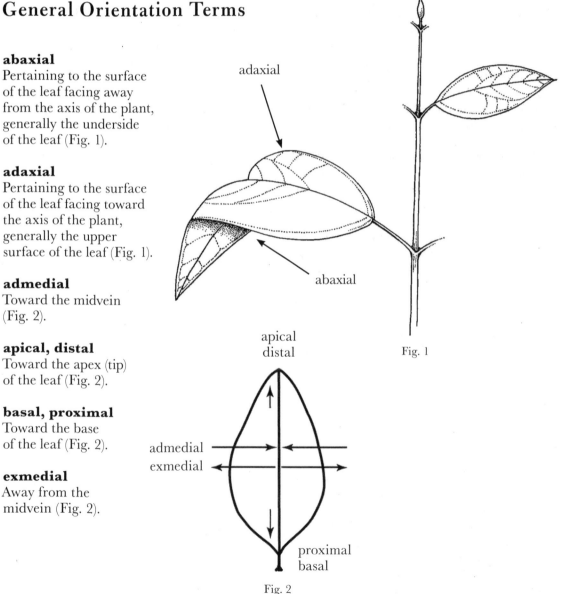

Fig. 1

Fig. 2

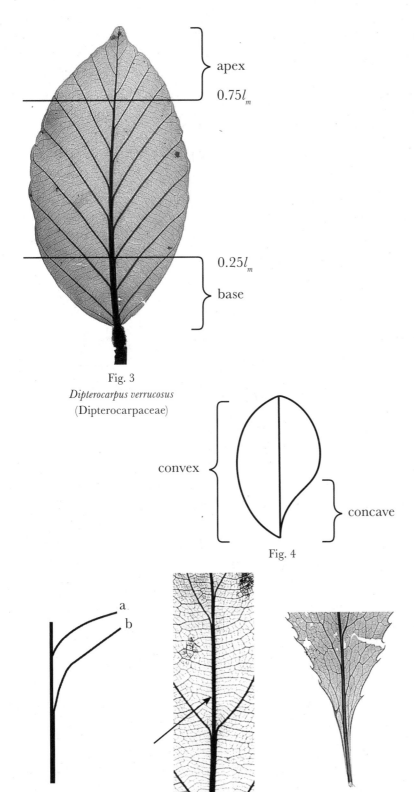

Fig. 3
Dipterocarpus verrucosus
(Dipterocarpaceae)

Fig. 4

Fig. 5

Fig. 6
Itea chinensis
(Iteaceae)

Fig. 7
Berberis sieboldii
(Berberidaceae)

apex

The distal ~25% of the lamina (Fig. 3). If the lamina has an apical extension (tissue distal to the point where the primary vein ends), the apex includes all tissue distal to 0.75 l_m, where l_m is the distance from the proximal to the distal end of the midvein. **Note:** See Figure 17 for a description of lamina length.

base

The proximal ~25% of the lamina (Fig. 3). If the lamina has a basal extension, the base includes all tissue proximal to 0.25 l_m, where l_m is the distance from the proximal to the distal end of the midvein **Note:** See Figure 17 for a description of lamina length.

concave

Curved inward relative to the midvein (Fig. 4).

convex

Curved outward relative to the midvein (Fig. 4).

decurrent

Approaching an intersection in an asymptotic manner in the basal direction (Fig. 5). Applies both to veins, as shown in Figure 6, and to laminar tissue, as shown in Figure 7. Note that decurrent secondary veins may simply branch, as shown in Figure 5a, or may "steal" part of the midvein, making the midvien thinner above the secondary, as shown in Figure 5b.

Parts of a Simple Leaf

lamina (blade)
The expanded, flattened portion of a leaf (Fig. 8).

leaf
The chief photosynthetic organ of most vascular land plants, usually a determinate outgrowth of a primordium produced laterally on an axis. Most leaves consist of a petiole (stalk), a leaf base, and a bifacial lamina (blade). Leaves subtend axillary buds and have a definite arrangement, or phyllotaxy, in their insertion along the axis (Fig. 8).

petiole
The stalk that attaches a leaf to the axis (Figs. 8, 10).

insertion point
The place where the base of the lamina joins the petiole (Fig. 9).

margin
The outer edge of the lamina (Fig. 10).

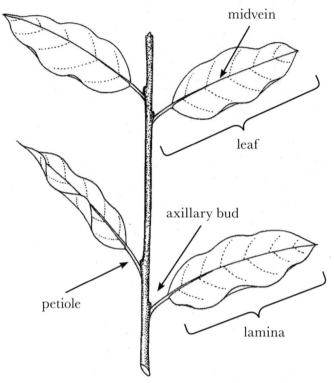

Fig 8

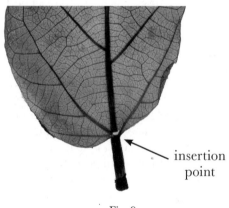

Fig. 9
Alangium chinense
(Cornaceae)

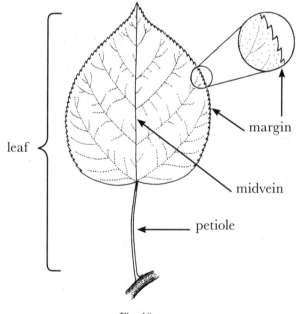

Fig. 10

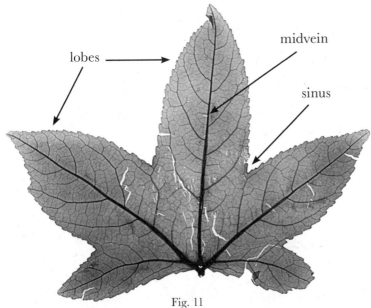

lobes

midvein

sinus

Fig. 11
Liquidambar styraciflua
(Altingiaceae)

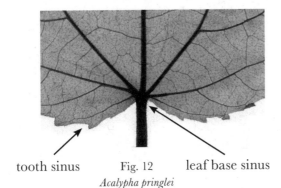

tooth sinus   Fig. 12   leaf base sinus
Acalypha pringlei
(Euphorbiaceae)

midvein
The medial primary vein. In pinnate leaves, it is the only primary vein (Figs. 8, 10, 11; see Section II, below, for further discussion of primary veins).

lobe
A marginal projection with a corresponding sinus incised 25% or more of the distance from the projection's apex to the midvein, measured parallel to the axis of symmetry and along the distal side of the projection or the basal side of a terminal projection (Fig. 11).

sinus
A marginal embayment, incision, or indentation between marginal projections of any sort, typically lobes (Fig. 11), teeth (Fig. 12), or the base of cordate leaves (Fig. 12).

leaf domatia
Cavities or hollow structures on the laminar, stipular, or petiolar surfaces of the leaf, inferred to be habitable by insects or mites (Fig. 13).

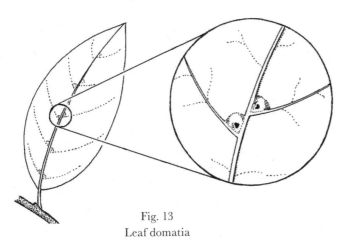

Fig. 13
Leaf domatia

Parts of a Compound Leaf

compound leaf
A leaf with two or more noncontiguous areas of laminar tissue (Fig. 14).

leaflet
A discrete, separate laminar segment of a compound leaf. Leaflets never subtend axillary buds (Fig. 14).

rachis
The prolongation of the petiole of a pinnately compound leaf, to which leaflets are attached (Fig. 14). In cross-section the rachis may be terete (round), semiterete, angular, canaliculate (having longitudinal channels), or winged (see Figs. 46–49 for the analogous characters in petioles). A second-order rachis is a *rachilla* (see Fig. 32).

petiolule
The stalk that attaches a leaflet of a compound leaf to its rachis (Fig. 14).

insertion point
The point where the leaf is attached to the axis or where a leaflet is attached to the petiole or petiolule (Fig. 14).

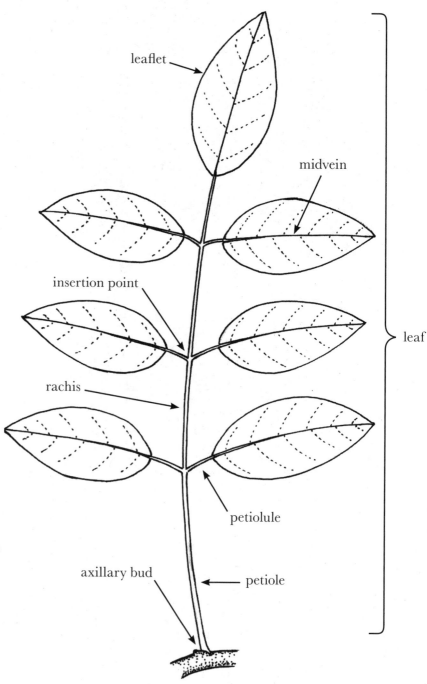

Fig. 14

Stipels and Stipules

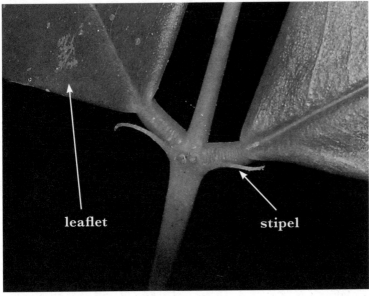

Fig. 15
Andira mandshurica
(Fabaceae)

Fig. 16
Malus baccata
(Rosaceae)

stipel
A stipule-like structure located at the base of the petiolule of some leaflets or extrafloral nectaries. Stipels may occur on the petiolule or at the juncture of the petiolule and rachis (Fig. 15).

stipule
On dicotyledonous plants, an outgrowth (scale, laminar structure, or spine) usually associated with the point of insertion of a leaf on a stem (Fig. 16). Stipules may occur on or along part of the base of the petiole but are more often on the axis near the petiole base, where they can be intrapetiolar (between petiole and stem), leaf-opposed, lateral, or (for opposite leaves) interpetiolar. They are usually paired but may be fused to form a single sheathing or perfoliate structure. Stipules are usually deciduous, often leaving behind a characteristic scar. Domatia, tendrils, or extrafloral nectaries may occupy stipule positions. Stipules may be difficult to distinguish from pseudostipules, stipule-like paired outgrowths on the petiole toward or rarely at the base of pinnately compound leaves that are morphologically distinct from the leaflets.

Measurements

lamina length, $L = l_m + l_a + l_b$ (Fig. 17).

apical extension length, l_a = Distance from the most distal point of the midvein to the most distal extension of leaf tissue, the latter projected to the trend of the midvein (Fig. 17c, d). In most leaves, $l_a = 0$ (Fig. 17a, b).

basal extension length, l_b = Distance from the most proximal point of the midvein to the most proximal extension of leaf tissue, the latter projected to the trend of the midvein (Fig. 17b, d; Fig. 18). In many leaves $l_b = 0$ (Fig. 17a, c). When l_b is longer on one side of the leaf than the other, always use the larger value when calculating lamina length (Fig. 18).

midvein length, l_m = Distance from the proximal end of the midvein to the distal end (Fig. 17).

width ratio, x/y = The ratio of the smaller to the larger of the two distances measured perpendicularly from the midvein to the margin on each side of the leaf at the position of maximum leaf width (Fig. 19). On a lobed leaf, the width ratio is measured to the outermost portion of the leaf (Fig. 20).

basal width ratio, similar to width ratio but measured only in the widest portion of the base of the leaf (Fig. 21).

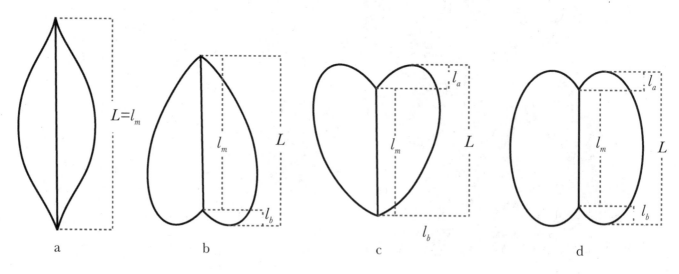

Fig. 17

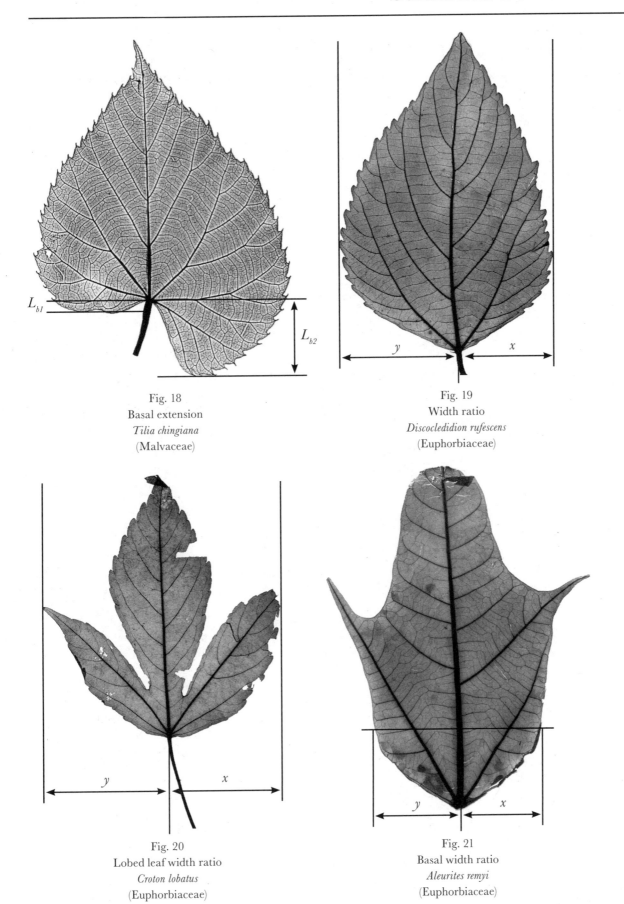

Fig. 18
Basal extension
Tilia chingiana
(Malvaceae)

Fig. 19
Width ratio
Discocledidion rufescens
(Euphorbiaceae)

Fig. 20
Lobed leaf width ratio
Croton lobatus
(Euphorbiaceae)

Fig. 21
Basal width ratio
Aleurites remyi
(Euphorbiaceae)

I. Leaf Characters

1. **Leaf Attachment**

    1.1 **Petiolate** – A petiole attaches the leaf to the axis (Figs. 8, 10, 13, 22).

    1.2 **Sessile** – Leaf attaches directly to the axis without a petiole (Fig. 23).

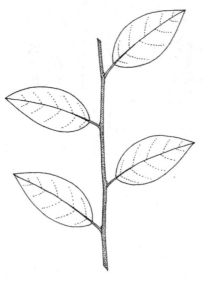

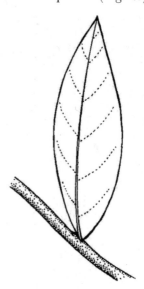

Fig. 22
Leaf attachment petiolate, leaf
arrangement alternate (distichous)

Fig. 23
Leaf attachment sessile

2. **Leaf Arrangement** – The placement of adjacent leaves on the nodes of the axis (more than one may apply). **Note:** For more detailed treatments of phyllotaxy, see Bell, 2008; or Keller, 2004.

    2.1 **Alternate** – Adjacent leaves occur above or below others on the axis with one leaf per node (Fig. 22). The arrangement may be distichous (in one plane in two ranks on opposite sides of the axis) or helical (in a spiral along the axis).

    2.2 **Subopposite** – Adjacent leaves occur in pairs that are nearly but not strictly opposite (Fig. 24). These pairs may be decussate (leaf pairs inserted at ~90° to those above and below), distichous (leaf pairs are aligned with those above and below), or spirodecussate (successive leaf pairs inserted at angles >90° to those above and below).

    2.3 **Opposite** – Leaves occur in opposed pairs that arise from the same node along the axis. Leaf pairs may be decussate (Fig. 25), distichous (Fig. 26), or spirodecussate.

    2.4 **Whorled** – Three or more leaves are borne at each node (Fig. 27).

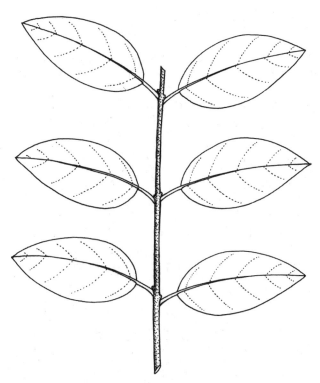

Fig. 24
Leaf arrangement subopposite (distichous)

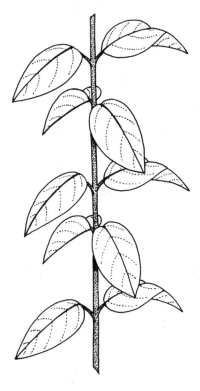

Fig. 25
Leaf arrangement opposite (decussate)

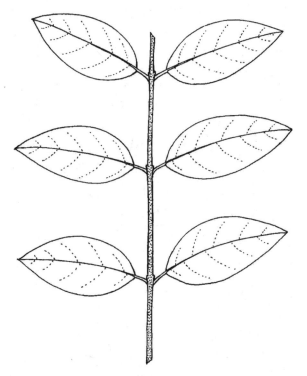

Fig. 26
Leaf arrangement opposite (distichous)

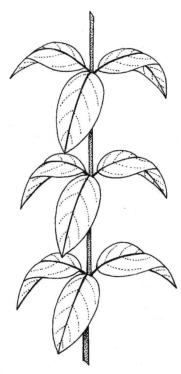

Fig. 27
Leaf arrangement whorled

3.   **Leaf Organization**

3.1   **Simple** – Leaf consists of a single lamina attached to a simple petiole (Fig. 28). This is the most common case.

3.2   **Compound** – Leaf consists of two or more leaflets (laminae not interconnected by laminar tissue.) **Note:** *Ternate,* a term used for various types of organization of leaflets (and leaves) into threes, is not treated here.

3.2.1   **Palmately compound** – Leaf has more than two separate laminar subunits (leaflets) attached at the apex of a petiole (Fig. 29). The description should include the number of leaflets.

3.2.2   **Pinnately compound** – Leaf has leaflets arranged along a rachis.

3.2.2.1   **Once compound** – With a single order of pinnate leaflets (Fig. 30, 31).

3.2.2.2   **Twice, or bipinnately compound** – Dissected twice with leaflets arranged along rachillae that are attached to the rachis (Fig. 32).

3.2.2.3   **Thrice, or tripinnately compound** – Leaflets are attached to secondary rachillae that are in turn attached to rachillae, which are borne along the rachis (Fig. 33).

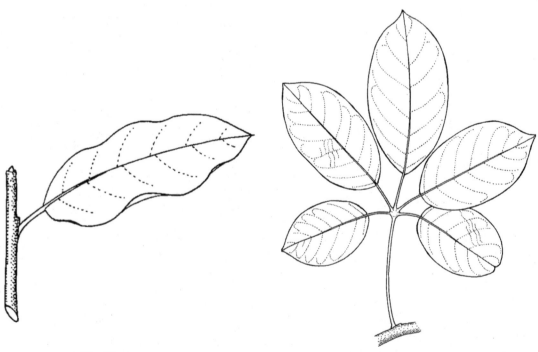

Fig. 28
Leaf organization simple

Fig. 29
Leaf organization palmately compound

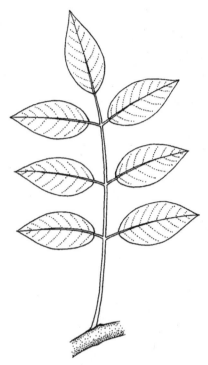

Fig. 30
Leaf organization once-pinnately compound (odd)

Fig. 31
Leaf organization once-pinnately compound (even)
Hymenaea courbaril
(Fabaceae)

Fig. 32
Leaf organization twice-pinnately compound

Fig. 33
Leaf organization thrice-pinnately compound

4.   **Leaflet Arrangement** – These character states apply only to pinnately compound leaves. Note that odd-pinnately compound (imparipinnate) leaves have a single terminal leaflet, and even-pinnately compound (paripinnate) leaves do not. These terms are illustrated for opposite leaflets but may apply to subopposite leaflets as well.

    4.1   **Alternate** – Leaflets are arranged alternately on the rachis (Fig. 34).

    4.2   **Subopposite** – Leaflets are in pairs that are nearly, but not strictly, opposite (Fig. 35).

    4.3   **Opposite** – Leaflets are in pairs that arise on opposite sides of the rachis.

        4.3.1   **Odd-pinnately compound** (Fig. 36).

        4.3.2   **Even-pinnately compound** (Fig. 37).

    4.4   **Unknown** – fossil only; not preserved (Fig. 38).

5.   **Leaflet Attachment** – These character states apply only to compound leaves.

    5.1   **Petiolulate** – Leaflet is attached to the rachis by means of a petiolule (stalk), analogous to the petiole of a leaf (Figs. 34–38).

    5.2   **Sessile** – Leaflet is attached directly to the rachis (Fig. 39).

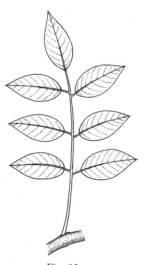

Fig. 34
Leaflet arrangement alternate; petiolulate

Fig. 35
Leaflet arrangement subopposite; petiolulate

Fig. 36
Leaflet arrangement opposite (odd)

Fig. 37
Leaflet arrangement
opposite (even)

Fig. 38
Leaflet arrangement
unknown

Fig. 39
Leaflet attachment
sessile

6.   Petiole Features

6.1   Petiole base

6.1.1   Sheathing – Petiole expands to clasp the stem (Fig. 40).

6.1.2   Pulvin(ul)ate – Having an abruptly swollen portion near the node around which the leaf(let) can flex (Fig. 41); may occur with or without an abscission joint (Fig. 42). On compound leaves, a pulvinulus may occur at the proximal and/or distal end of the petiolule and sometimes only on the terminal leaflet (Fig. 43).

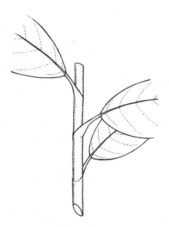

Fig. 40
Petiole base sheathing

Fig. 41
Petiole base pulvinate
Antrocaryon amazonicum
(Anacardiaceae)

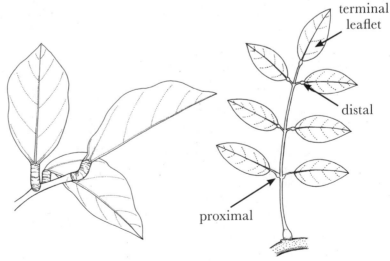

Fig. 42
Petiolule base pulvinulate

Fig. 43
Position of pulvinulus

6.2   **Glands** (see also I.22 and III.53) – Swollen areas of secretory tissue, often paired.

   6.2.1   **Petiolar** – Glands are borne along the petiole (Fig. 44).

   6.2.2   **Acropetiolar** – Glands are borne at the distal end of the petiole, below the base of the leaf (Fig. 45).

6.3   **Petiole-cross section**

   6.3.1   **Terete** – Round (Fig. 46).

   6.3.2   **Semiterete** – Semicircular (Fig. 47).

   6.3.3   **Canaliculate** – Having a longitudinal channel or groove (Fig. 48).

   6.3.4   **Angular** (Fig. 49).

   6.3.5   **Alate or Winged** – With lateral ridges or flanked by laminar tissue (Fig. 50).

6.4   **Phyllodes** – Petiole or rachis is expanded to make a lamina (Fig. 51).

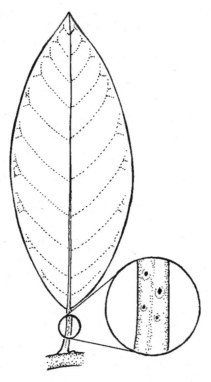

Fig. 44
Petiolar glands

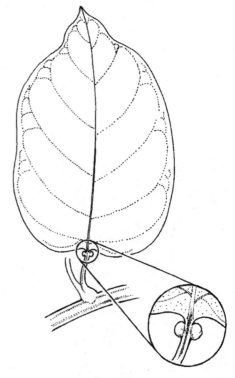

Fig. 45
Acropetiolar glands

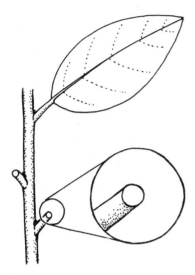

Fig. 46
Terete petiole cross-section

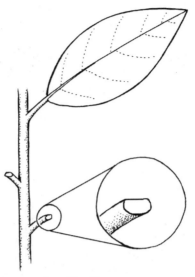

Fig. 47
Semi-terete petiole cross-section

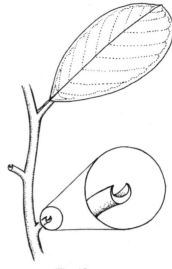

Fig. 48
Canaliculate petiole cross-section

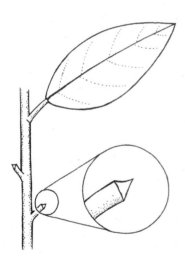

Fig. 49
Angular petiole cross-section

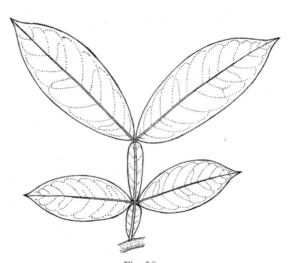

Fig. 50
Alate petiole and rachis

Fig. 51
Phyllode
Acacia mangium
(Fabaceae-Mimosoideae)

7. **Position of Lamina Attachment** – The point from which the lamina is borne.

    7.1   **Marginal** – Leaf is attached at its margin (Fig. 52).

    7.2   **Peltate central** – Leaf is borne from a position near the center of the lamina (Fig. 53).

    7.3   **Peltate excentric** – Leaf is borne from a position within the boundaries of the lamina but not near its center (Fig. 54).

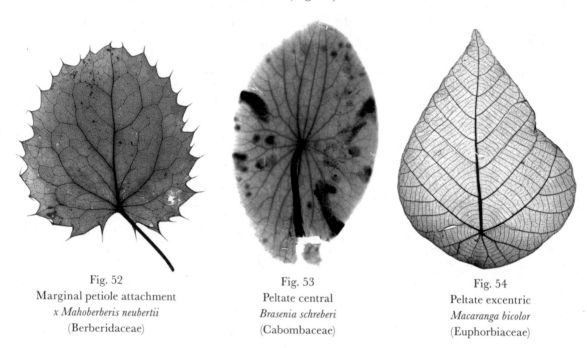

Fig. 52	Fig. 53	Fig. 54
Marginal petiole attachment	Peltate central	Peltate excentric
x Mahoberberis neubertii	*Brasenia schreberi*	*Macaranga bicolor*
(Berberidaceae)	(Cabombaceae)	(Euphorbiaceae)

8. **Laminar Size** – The area of the leaf blade. When possible, the area should be measured directly (e.g., digitally) or approximated by multiplying the length by the width by 0.75 (Cain and Castro, 1959). Alternatively, laminar size can be approximated by size classes (Raunkiaer, 1934; Webb, 1959). Figure 55 shows outlines of the maximum sizes of five of the smallest size classes; the leaf belongs in the smallest size class into which its area fits completely. The template, which can be photocopied onto clear acetate and placed over a leaf, is included for paleobotanists, who often work with incomplete fossil leaves and must approximate leaf area.

    Areas of leaf size classes (Webb, 1959):

8.1	**Leptophyll**	<25 mm^2
8.2	**Nanophyll**	25–225 mm^2
8.3	**Microphyll**	225–2,025 mm^2
8.4	**Notophyll**	2,025–4,500 mm^2
8.5	**Mesophyll**	4,500–18,225 mm^2
8.6	**Macrophyll**	18,225–164,025 mm^2
8.7	**Megaphyll**	$>164,025$ mm^2

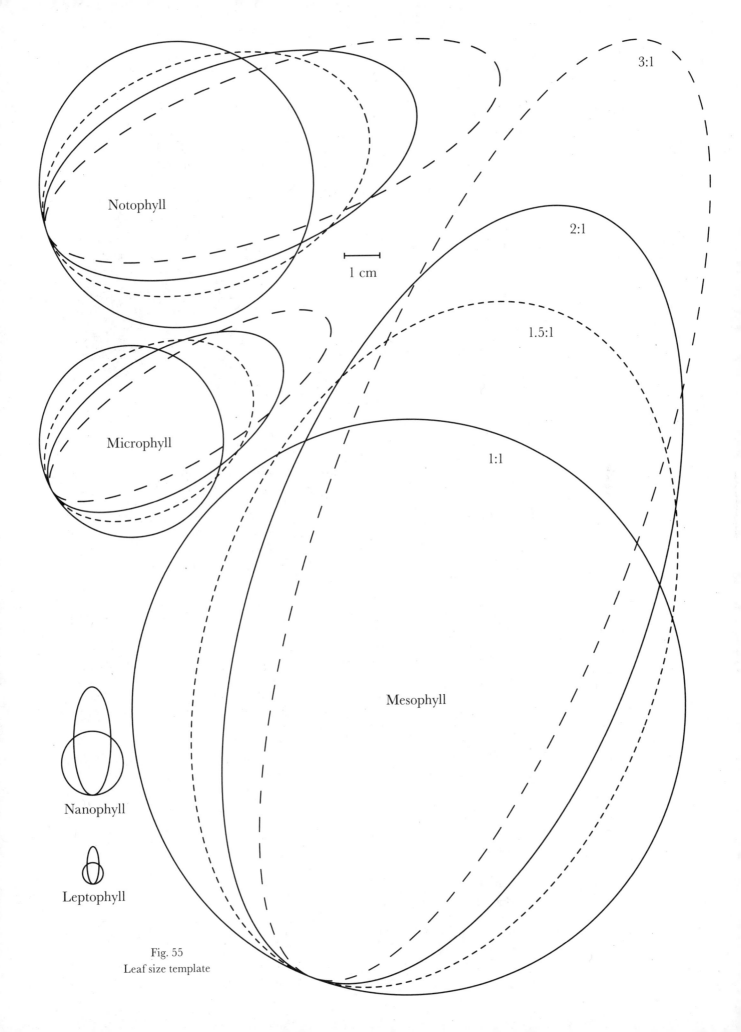

Notophyll

3:1

2:1

1.5:1

1 cm

Microphyll

1:1

Mesophyll

Nanophyll

Leptophyll

Fig. 55
Leaf size template

9. **Laminar L:W Ratio** – Ratio of laminar length to maximum width perpendicular to the axis of the midvein (Fig. 56).

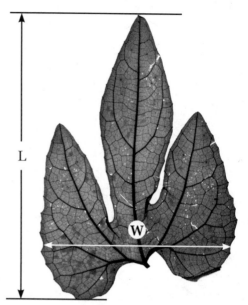

Fig. 56
Trichosanthes formosana
(Curcurbitaceae)

10. **Laminar Shape (in compound leaves, this applies to the shape of the leaflets)** – To determine the shape of the lamina, locate the midvein and determine the zone of greatest width measured perpendicular to the midvein. In lobed leaves, draw a line from the apex to the widest point on either side of the midvein and determine the shape by finding the zone of greatest width based on this outline (Fig. 57). Historically, botanists combined leaf shape with imprecisely defined L:W ratios to create additional character states (e.g., von Ettingshausen, 1861). Some common historical terms are italicized below but not illustrated.

   10.1 **Elliptic** – The widest part of the leaf is in the middle one-fifth (Fig. 58). **Note:** The terms *orbiculate* and *oblate* have been used to describe unlobed, elliptic leaves that are very wide. We suggest using *orbiculate* for elliptic leaves with a L:W ratio ranging from 1.2:1 to 1:1 and *oblate* for elliptic leaves with a L:W ratio <1:1.

   10.2 **Obovate** – The widest part of the leaf is in the distal two-fifths (Fig. 59). We suggest defining *oblanceolate* leaves as obovate leaves with a L:W ratio between 3:1 and 10:1.

   10.3 **Ovate** – The widest part of the leaf is in the proximal two-fifths (Fig. 60). **Note:** *Lanceolate* has been used to describe ovate leaves that are long and narrow. We suggest defining *lanceolate* leaves as ovate leaves with a L:W ratio between 3:1 and 10:1.

   10.4 **Oblong** – The opposite margins are roughly parallel for at least the middle one-third of the leaf (Fig. 61).

   10.5 **Linear** – The L:W ratio of a leaf is ≥10:1, regardless of the position of the widest part of the leaf (Fig. 62).

   10.6 **Special** – Outlines that do not fall readily into one of the shape classes above; for example, the pitcher-shaped leaf apex of *Nepenthes*.

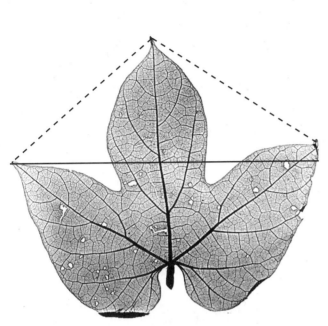

Fig. 57
Measuring lobed leaves
Dioscoreophyllum strigosum
(Menispermaceae)

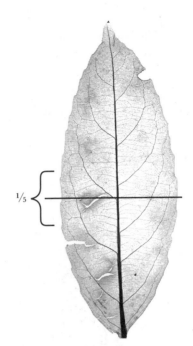

Fig. 58
Elliptic leaf shape
Cheiloclinium anomalum
(Celastraceae)

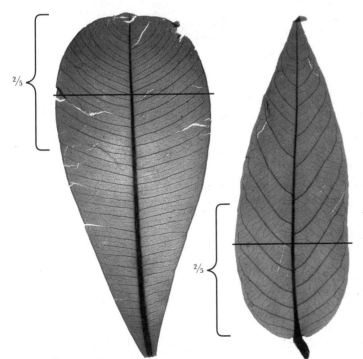

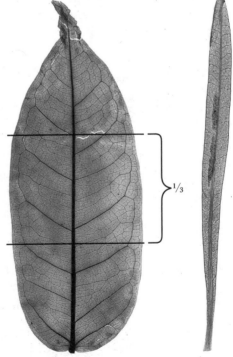

Fig. 59
Obovate leaf shape
Alstonia congensis
(Apocynaceae)

Fig. 60
Ovate leaf shape
Parinari sp.
(Chrysobalancaceae)

Fig. 61
Oblong leaf shape
Ficus citrifolia
(Moraceae)

Fig. 62
Linear leaf shape
Xylomelum angustifolium
(Proteaceae)

11.   **Medial Symmetry** – Determined by the width ratio in the middle of the leaf (see Measurements, above).

    11.1   **Symmetrical** – Width ratio (x/y) > 0.9 from 0.25L to 0.75L (Fig. 63).

    11.2   **Asymmetrical** – Width ratio (x/y) < 0.9 from 0.25L to 0.75L (Fig. 64).

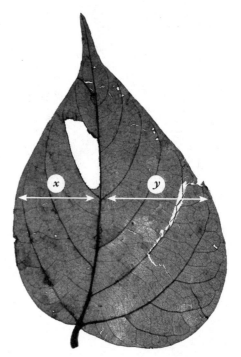

<table>
<tr><td>Fig. 63
Leaf medially symmetrical
Maytenus aquifolium
(Celastraceae)</td><td>Fig. 64
Leaf medially asymmetrical
Ramirezella pringlei
(Fabaceae)</td></tr>
</table>

12.   **Base Symmetry** – Base symmetry and basal width asymmetry are determined by the width ratio in the base of the leaf (see Measurements, above). Leaf bases can be asymmetrical in insertion, extension, and width.

    12.1   **Base Symmetrical** – Base lacks any of the asymmetries identified below (Fig. 65).

    12.2   **Base Asymmetrical**

        12.2.1   **Basal width asymmetrical** – Basal width ratio (x/y) < 0.9 (Fig. 66).

        12.2.2   **Basal extension asymmetrical** – Basal extension length on one side is <0.75 of the other side (L_{b1}/L_{b2} < 0.75) (Fig. 67).

        12.2.3   **Basal insertion asymmetrical** – Insertion points of lamina base on either side of the petiole are separated by >3 mm (Fig. 68).

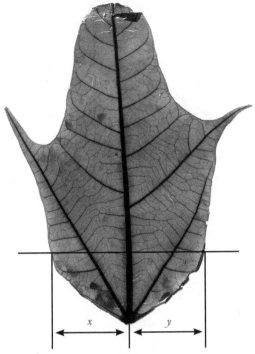

Fig. 65
Basal width symmetrical
Aleurites remyi
(Euphorbiaceae)

Fig. 66
Basal width asymmetrical
Lunania mexicana
(Salicaceae)

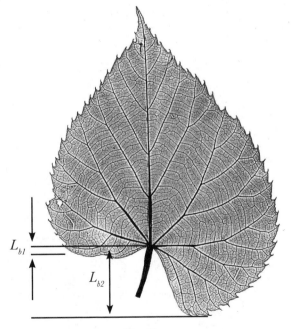

Fig. 67
Basal extension asymmetrical
Tilia chingiana
(Malvaceae)

Fig. 68
Basal insertion asymmetrical
Fraxinus floribunda
(Oleaceae)

13.   **Lobation** – A lobe is a marginal projection with a corresponding sinus incised 25% or more of the distance from the projection apex to the midvein, measured parallel to the axis of symmetry and along the apical side of the projection (or the basal side of a terminal projection). A leaf is considered lobed even if it has only one marginal projection that fits the definition. If the sinus described above is incised less than 25% of the width, the projection is considered a tooth (see Section III).

13.1   **Unlobed** – The leaf has no lobes (Figs. 69, 70). Note that the leaf in Figure 69 is also called "entire" because it lacks lobes and teeth. The term *entire* is useful because it describes the majority of angiosperm leaves. For further discussion of *entire*, see I.14.

13.2   **Lobed**

13.2.1   **Palmately lobed** – Major veins of the lobes are primary veins that arise from the base of the leaf (Fig. 71).

13.2.1.1 **Palmatisect** – Special case of palmately lobed in which the incision goes almost to the petiole but without resulting in distinct leaflets (Fig. 72). *Palmatifid* and *palmatipartite* are variously used terms for leaves with incised palmate lobes that are not treated here.

13.2.2   **Pinnately lobed** – Major veins of the lobes are formed by costal secondaries (Fig. 73).

13.2.2.1 **Pinnatisect** – Special case of pinnately lobed in which the incision goes almost to the midvein but without resulting in distinct leaflets (Fig. 74). *Pinnatifid* and *pinnatipartite* are variously used terms for leaves with pinnately-incised lobes that are not treated here.

13.2.3   **Palmately and pinnately lobed** – At least one lobe in a palmately lobed leaf is pinnately lobed (Fig. 75).

13.2.4   **Bilobed** – Leaf has two lobes (Fig. 76).

Fig. 69
Unlobed (entire)
Parinari campestris
(Chrysobalanaceae)

Fig. 70
Unlobed (with teeth)
Melanolepis multiglandulosa
(Euphorbiaceae)

Fig. 71
Palmately lobed
Adenia heterophylla
(Passifloraceae)

Fig. 72
Palmatisect
Potentilla recta
(Rosaceae)

Fig. 73
Pinnately lobed
Stenocarpus sinuatus
(Proteaceae)

Fig. 74
Pinnatisect
Dryandra longifolia
(Proteaceae)

Fig. 75
Palmately and pinnately lobed
Cucurbita cylindrata
(Curcurbitaceae)

Fig. 76
Bilobed
Bauhinia madagascariensis
(Fabaceae)

14.   **Margin Type** – Features of the edge of the lamina. Section I.13 describes how to distinguish lobes and teeth.

14.1   **Untoothed** – Margin has no teeth (Fig. 77). **Note:** The term *entire* describes a leaf with no teeth and no lobes (Fig. 69). Leaf Margin Analysis and other physiognomic methods of paleoclimate inference score lobed leaves without teeth in the same category as entire leaves (Wolfe 1995), thus the category "entire" has sometimes been inferred to include lobed, untoothed leaves. We prefer the word *untoothed* for this category because it provides the clearest alternative to *toothed* and does not conflict with the standard botanical meaning of *entire*, which excludes all lobed leaves.

14.2   **Toothed** – Margin has vascularized projections (Figs. 78–80) separated by sinuses that are incised less than 25% of the distance to the midvein or long axis of the leaf as measured parallel to the axis of symmetry from the apical incision of the projection. Note that a leaf with a single tooth of any size is considered toothed. Also, both lobes and teeth may be present on the same leaf (but see notes below).

14.2.1   **Dentate** – Majority of the teeth have axes of symmetry directed perpendicular to the trend of the leaf margin (Fig. 78).

14.2.2   **Serrate** – Majority of the teeth have axes of symmetry directed at an angle to the trend of the leaf margin (Fig. 79).

14.2.3   **Crenate** – Majority of the teeth are smoothly rounded, without a pointed apex (Fig. 80). **Note:** Crenate margins are also either dentate or serrate.

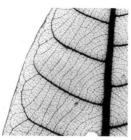

Fig. 77
Untoothed margin
Caraipa punctulata
(Clusiaceae)

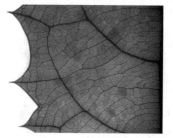

Fig. 78
Dentate margin
Casearia ilicifolia
(Salicaceae)

Fig. 79
Serrate margin
Betula lenta
(Betulaceae)

Fig. 80
Crenate and serrate margin
Viola brevistipulata
(Violaceae)

Notes: The difference between lobes and teeth is sometimes ambiguous. Some leaves have geometrically similar projections that could be scored as lobes or teeth using the 25% rule above. When at least one definitive lobe is present, we suggest scoring such projections as lobes and not as teeth (Fig. 81) (Royer et al., 2005). Some toothed leaves have projections at the apex that are incised more than 25%. We suggest scoring these projections as teeth rather than lobes (Fig. 82).

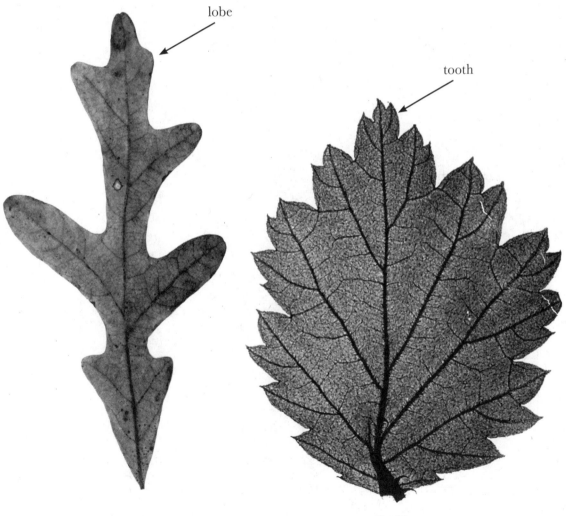

Fig. 81
Lobe that looks like a tooth
Quercus alba
(Fagaceae)

Fig. 82
Tooth that looks like a lobe
Rubus mesogaeus
(Rosaceae)

15. **Special Margin Features**

15.1 **Appearance of the edge of the leaf blade**

15.1.1 **Erose** – Margin is minutely irregular, as if chewed (Fig. 83).

15.1.2 **Sinuous** – Margin forms a series of shallow and gentle curves that lack principal veins. These projections are not considered teeth (see above or Section III) (Fig. 84).

15.2 **Appearance of the abaxial-adaxial plane of the leaf blade**

15.2.1 **Revolute** – Margin is turned down or rolled (in the manner of a scroll) in the abaxial direction (Fig. 85).

15.2.2 **Involute** – Margin is turned up or rolled in the adaxial direction (Fig. 86).

15.2.3 **Undulate** – Margin forms a series of smooth curves in the abaxial-adaxial plane (in and out of the plane of the leaf) (Fig. 87).

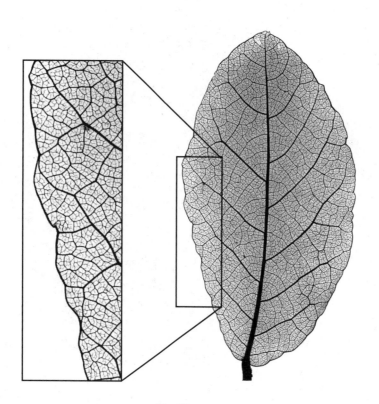

Fig. 83
Erose margin
Bridelia cathartica
(Phyllanthaceae)

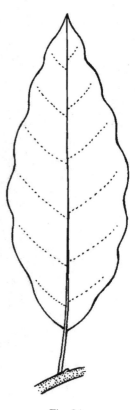

Fig. 84
Sinuous margin

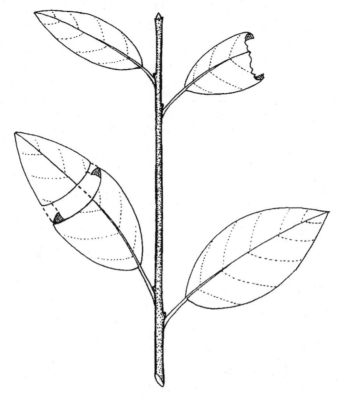

Fig. 85
Revolute margin

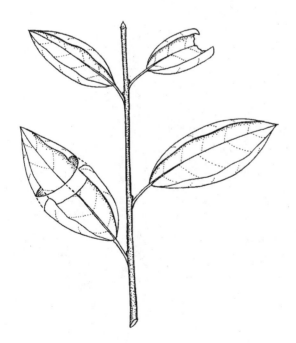

Fig. 86
Involute margin

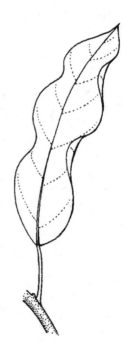

Fig. 87
Undulate margin

16.   Apex Angle

The vertex of the apex angle lies at the center of the midvein where it terminates at the apex of the leaf. The apex angle is formed by the two rays that depart this vertex and are tangent to the leaf margin without crossing over any part of the lamina (Figs. 88, 89). The apex angle is always measured on the proximal side of the rays. If the leaf is toothed, draw the lines along the edge of the margin, connecting the marginal tissue (Fig. 89). If the midvein terminates between two lobes, the angle is formed as in unlobed leaves but is greater than 180° (Fig. 90). If the midvein terminates at the apex of a lobe, the rays need only be tangent to the margin of the terminal lobe and may pass over lateral lobes (Fig. 91). Leaves with retuse apices (see 20.3) are considered to have an obtuse apex angle. The following categories are useful for scoring apex angles:

16.1   **Acute** – Apex angle <90° (Fig. 88).

16.2   **Obtuse** – Apex angle between 90° and 180° (Fig. 89).

16.3   **Reflex** – Apex angle >180° (Fig. 90).

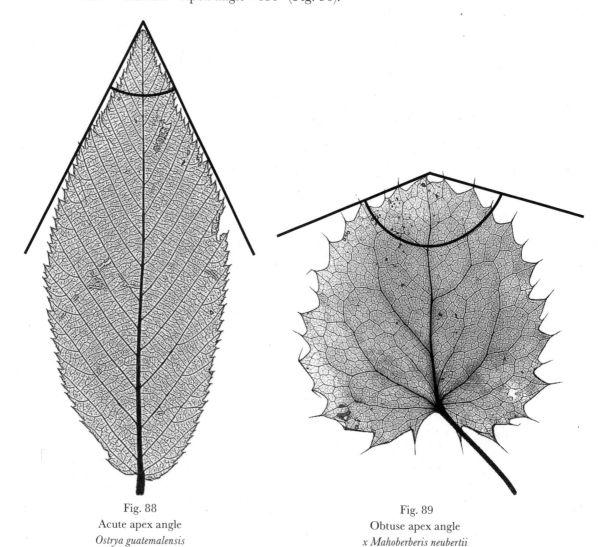

Fig. 88
Acute apex angle
Ostrya guatemalensis
(Betulaceae)

Fig. 89
Obtuse apex angle
x Mahoberberis neubertii
(Berberidaceae)

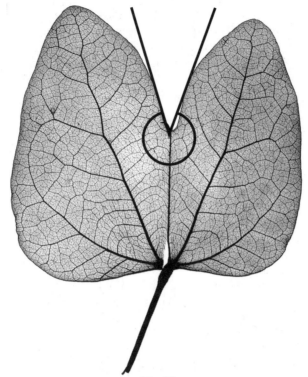

Fig. 90
Reflex apex angle
Bauhinia madagascariensis
(Fabaceae)

Fig. 91
Acute apex angle on a lobed leaf
Dioscoreophyllum strigosum
(Menispermaceae)

17.   **Apex Shape** – These states apply to the shape of the distal 25% of the lamina. On a toothed leaf, a smoothed curve through the tips of the teeth determines the shape (Fig. 93). For leaves with an apical extension ($l_a > 0$), follow the guidelines in Figure 92. If the apex is retuse (see also 20.3), it can still be scored for the other shape features given below.

17.1   **Straight** – Margin between the apex and 0.75L has no significant curvature (Fig. 93).

17.2   **Convex** – Margin between the apex and 0.75L curves away from the midvein (Fig. 94).

17.2.1   **Rounded** – Subtype of convex in which the margin forms a smooth arc across the apex (Fig. 95).

17.2.2   **Truncate** – Apex terminates abruptly as if cut, with margin perpendicular to midvein or nearly so (Fig. 96).

17.3   **Acuminate** – Margin between the apex and 0.75L is convex proximally and concave distally, or concave only. This category, especially when the distal portion of the apex abruptly narrows, accommodates most apex types called "drip tips" (Figs. 97, 98).

17.4   **Emarginate** – l_m is 75–95% of $l_m + l_a$ (Fig. 99); see also *retuse* (20.3).

17.5   **Lobed** – l_m is <75% of $l_m + l_a$ (Fig. 90).

Note: If the leaf has a different apex shape on either side, both shapes should be recorded (Fig. 100).

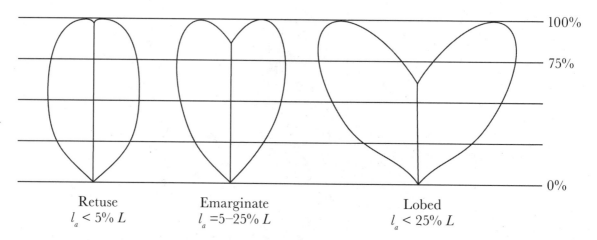

Retuse
$l_a < 5\% L$

Emarginate
$l_a = 5–25\% L$

Lobed
$l_a < 25\% L$

Fig. 92
Definitions of apex shapes for leaves that have an apical extension

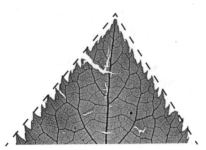

Fig. 93
Apex shape straight
Aristotelia racemosa
(Elaeocarpaceae)

Fig. 94
Apex shape convex
Saurauia calyptrata
(Actinidiaceae)

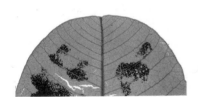

Fig. 95
Apex shape rounded
Ozoroa obovata
(Anacardiaceae)

Fig. 96
Apex shape truncate
Liriodendron chinense
(Magnoliaceae)

Fig. 97
Apex shape acuminate (with drip tip)
Neouvaria acuminatissima
(Annonaceae)

Fig. 98
Apex shape acuminate (without drip tip)
Corylopsis veitchiana
(Hamamelidaceae)

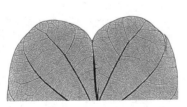

Fig. 99
Apex shape emarginate
Lundia spruceana
(Bignoniaceae)

Fig. 100
Apex shape acuminate on the left
and straight on the right
Tapura guianensis
(Dichapetalaceae)

18.   **Base Angle** – The vertex of the base angle lies in the center of the midvein next to the point where the basalmost laminar tissue joins the petiole (or joins the proximal margin in the case of sessile leaves). The base angle is formed by the two rays that depart this vertex and are tangent to the leaf margin without crossing over any part of the lamina. The base angle is independent of base shape (see Base Shape, I.19).

For consistency, the base angle is always measured on the distal side of the vertex, even when the angle is greater than 180° (Fig. 103–104). The following categories are useful for scoring base angles

18.1   **Acute** – Angle <90° (Fig. 101).

18.2   **Obtuse** – Angle >90° but <180° (Fig. 102).

18.3   **Reflex** – Special case of obtuse in which angle is >180° but <360° (Figs. 103, 104).

18.4   **Circular** – Special case of reflex in which angle is >360°. This includes leaves in which the basal extension overlaps across the midline, as well as peltate leaves (Fig. 105).

Fig. 101
Acute base angle
Schumacheria castaneifolia
(Dilleniaceae)

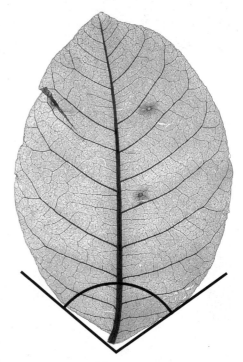

Fig. 102
Obtuse base angle
Mauria heterophylla
(Anacardiaceae)

Fig. 103
Reflex base angle
Tilia chingiana
(Malvaceae)

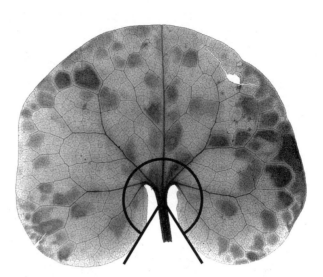

Fig. 104
Reflex base angle
Asarum europaeum
(Aristolochiaceae)

Fig. 105
Circular base angle
Cissampelos owariensis
(Menispermaceae)

19.   **Base Shape** – These states apply to the shapes of the proximal 25% of the lamina. On a toothed leaf, a smoothed curve through the tips of the teeth determines the shape.

19.1   **If there is no basal extension ($l_b = 0$), the following base types are recognized**

19.1.1   **Straight (cuneate)** – Margin between the base and 0.25L has no significant curvature (Fig. 106).

19.1.2   **Concave** – Margin between the base and 0.25L curves toward the midvein (Fig. 107).

19.1.3   **Convex** – Margin between the base and 0.25L curves away from the midvein (Fig. 108).

19.1.3.1 **Rounded** – The margin forms a smooth arc across the base (Fig. 109).

19.1.3.2 **Truncate** – The base terminates abruptly as if cut perpendicular to the midvein or nearly so (Fig. 110).

19.1.4   **Concavo-convex** – Margin between the base and 0.25L is concave proximally and convex distally (Fig. 111).

19.1.5   **Complex** – Margin curvature has more than one inflection point (change of curvature) between the base and 0.25L (Fig. 112).

19.1.6   **Decurrent** – Special case in which the laminar tissue extends along the petiole at a gradually decreasing angle (Figs. 113, 114); can occur in concave, concavo-convex, or complex bases.

19.2   **If there is a basal extension ($l_b > 0$), the following base types are recognized**

19.2.1   **Cordate** – Leaf base forms a single sinus with the petiole generally inserted at the deepest point of the sinus (Figs. 115, 116).

19.2.2   **Lobate** – Leaf base is lobed on both sides of the midvein. The lobes are defined by a central sinus containing the petiole as in cordate leaves, and by sinuses on their distal sides such that the nadirs of the distal sinuses are within the base of the leaf (Figs. 117, 118). The following terms have been used historically for some leaves that have two basal projections. We consider them to be subtypes of lobate bases.

19.2.2.1 **Sagittate** – Leaf base has two narrow, usually pointed projections (technically these may not qualify as lobes because they are not bounded by distal sinuses) with apices directed proximally at an angle 125° or greater from the midvein (Fig. 119).

19.2.2.2 **Hastate** – Leaf base has two narrow lobes with apices directed exmedially at 90°–125° from the midvein (Fig. 120).

19.2.2.3 Runcinate (not pictured) – A lobate lamina with two or more pairs of downward-pointing (>110°) angular lobes.

19.2.2.4 Auriculate (not pictured) – A lobate lamina having a pair of rounded basal lobes that are oriented downward, with their axes of symmetry at an angle >125° from the midvein of the leaf. If the lateral sinuses that define the lobes extend more than 50% of the distance to the midvein, such laminar bases may be referred to as *panduriform*.

Fig. 106
Base shape cuneate
Carya leiodermis
(Juglandaceae)

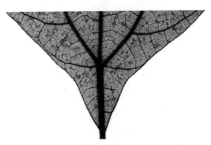

Fig. 107
Base shape concave
Sassafras albidum
(Lauraceae)

Fig. 108
Base shape convex
Prunus mandshurica
(Rosaceae)

Fig. 109
Base shape rounded
Carissa opaca
(Apocynaceae)

Fig. 110
Base shape truncate
Populus dimorpha
(Salicaceae)

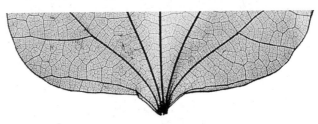

Fig. 111
Base shape concavo-convex
Diploclisia chinensis
(Menispermaceae)

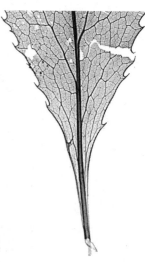

Fig. 112
Base shape complex
Adelia triloba
(Euphorbiaceae)

Fig. 113
Base shape decurrent
Alstonia plumosa
(Apocynaceae)

Fig. 114
Base shape decurrent
Berberis sieboldii
(Berberidaceae)

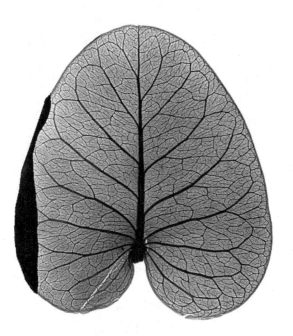

Fig. 115
Base shape cordate
Phyllanthus poumensis
(Phyllanthaceae)

Fig. 116
Base shape cordate
Cercidiphyllum japonicum
(Cercidiphyllaceae)

Fig. 117
Base shape lobate
Acer saccharinum
(Sapindaceae)

Fig. 118
Base shape lobate
Liquidambar styraciflua
(Hamamelidaceae)

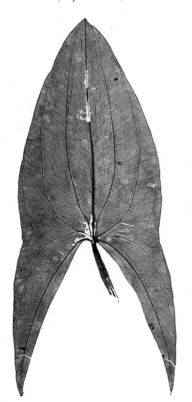

Fig. 119
Base shape sagittate
Sagittaria sp.
(Alismataceae)

Fig. 120
Base shape hastate
Araujia angustifolia
(Apocynaceae)

20. **Terminal Apex Features** – The following characters describe the region where the midvein terminates.

   20.1 **Mucronate** (apiculate) – The midvein terminates in an opaque, peg-shaped, nondeciduous extension of the midvein (Fig. 121).

   20.2 **Spinose** – The midvein extends through the margin at the apex; the spine may be short or long, but it is not always sharp (Fig. 122).

   20.3 **Retuse** – The midvein terminates in a shallow sinus such that l_m is 95–99% of $l_m + l_a$ (Fig. 123).

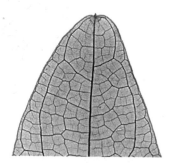

Fig. 121
Terminal apex mucronate
Cocculus ferrandianus
(Menispermaceae)

Fig. 122
Terminal apex spinose
Bauhinia rubeleruziana
(Fabaceae)

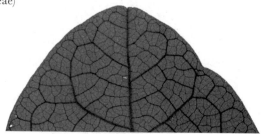

Fig. 123
Terminal apex retuse
Fitzalania heteropetala
(Annonaceae)

21. **Surface Texture** (see Stearn, 1983)

   21.1 **Smooth** – Lacking indentations, projections, hairs, or other roughness.

   21.2 **Pitted** – Having indentations.

   21.3 **Papillate** – Having small projections originating from the laminar surface.

   21.4 **Rugose** – Rough; for example, from vein relief.

   21.5 **Pubescent** – Having hairs (see Theobald et al., 1979, or Hewson, 1988, for pubescence categories).

22.   **Surficial Glands** – Placement of secretory structures.

    22.1   **Laminar** – Glands present on the surface (may be clustered) (Fig. 124).

    22.2   **Marginal** – Glands present only near or on the blade margin (Fig. 125).

    22.3   **Apical** – Glands present only near the blade apex (Fig. 126).

    22.4   **Basilaminar** – Glands present only near the base of the blade (Fig. 127).

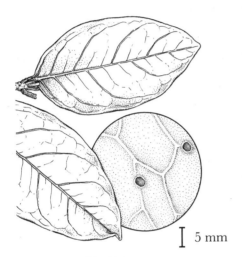

Fig. 124
Surficial glands laminar

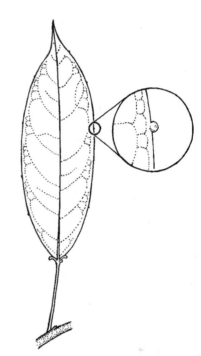

Fig. 125
Surficial glands marginal

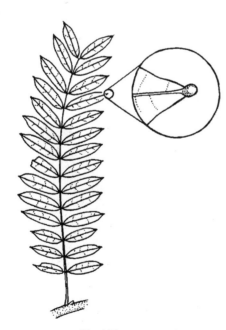

Fig. 126
Surficial glands apical

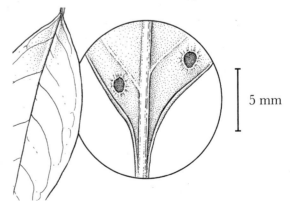

Fig. 127
Surficial glands basilaminar

ramified
Branching into higher-order veins without rejoining veins of the same or lower orders (Fig. 136).

sympodial
Type of branching in which the main vein axis is deflected at each branch point (Figs. 137, 138).

vein course
The path of a vein.

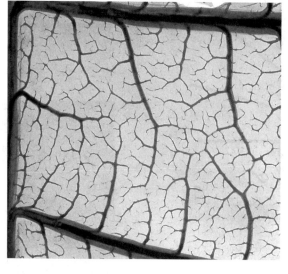

Fig. 136
Ramified veins
Comocladia cuneata
(Anacardiaceae)

Fig. 137
Sympodial branching

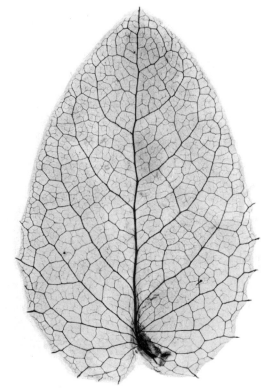

Fig. 138
Sympodial branching of primary
Griselinia scandens
(Griseliniaceae)

Determining Vein Order and Type

The first step in describing the pattern of venation in a leaf is to recognize categories, or *orders*, of veins that have visually distinct gauges and courses. Most angiosperm leaves have between four and seven orders of venation, which are conventionally numbered sequentially, starting with 1° for the primary vein or veins.

In general, the primary (1°) and secondary (2°) veins are the major structural veins of the leaf, and the tertiary (3°) veins are the largest-gauge veins that constitute the mesh, or "fabric," of the vein system. The primary vein or veins are somewhat analogous to the main trunk or trunks of a tree—they have the largest gauge, they usually taper along their length, and they generally run from the base or near the base of the leaf to its margin at the apex. Secondary veins are analogous to the major limbs of a tree. They are the next set in gauge after the primary or primaries, they also usually taper along their course, and they ordinarily run either from the base of the leaf or from a primary vein toward the margin. The tree analogy breaks down for 3° and higher-order veins because these veins maintain a similar gauge along their courses, and because they may form a reticulum, or net.

Tertiary veins usually have a narrower gauge than the secondary set, have courses that often connect 1° and 2° veins to each other throughout the leaf, and are the veins of highest gauge that form a more or less organized "field" over the great majority of the leaf area. Generally,

it is fairly easy to recognize the primaries and tertiaries, but the secondaries sometimes consist of several subsets with different gauges and courses. Nevertheless, all the subsets of veins between the primaries and the tertiaries are considered to be secondaries.

After the three lowest vein orders have been demarcated, the higher orders of venation (4°–7°) present in the leaf can be identified. Each of these higher vein orders may be highly variable among species and higher taxa in its degree of distinctness from both the next higher and next lower vein orders. This may be true even within a single leaf. Good diagnostic features for distinguishing higher orders of veins from one another are (1) excurrent origin from their source veins and (2) a distinctly narrower gauge. If they arise dichotomously or appear to have the same, or nearly the same, gauge as their parent vein, they are considered the same order as the source vein.

The simultaneous use of two criteria for determining vein order introduces a degree of ambiguity into the process, because some veins may have the gauge typical of one vein order but the course typical of a different vein order. On the other hand, recognizing orders based solely on their gauge or solely on their course leads to illogical situations in which veins that appear to have different functions and developmental origins are assigned to the same order. Assigning veins to orders also has a somewhat arbitrary aspect because variations in gauge and course are not mathematically discrete (Bohn et al., 2002); for example, a vein may be intermediate in gauge between the 1° vein and the 2° veins. Natural breaks in gauge usually occur at vein branching points, however, so most veins can be assigned to an order unambiguously using visual cues.

The regularity of vein systems varies widely, but it can be described semi-quantitatively in terms of "leaf rank" (Hickey, 1977). Leaf rank has practical significance for recognizing vein orders because vein systems that are less well organized (i.e., have lower rank) also tend to have less distinct vein orders. Even 2° and 3°

veins may be difficult to distinguish in leaves of low rank (Fig. 139).

In our experience, different observers following a consistent set of rules can usually define vein orders in a repeatable manner for a given leaf (Figs. 140, 141). It is generally good practice to discriminate vein orders at the point where they are expressed at their widest gauge, usually nearest to the center or base of the leaf. The following is a set of guidelines for recognizing vein orders; see also the definitions that follow.

Fig. 139
Delphinium cashmerianum
(Ranunculaceae)

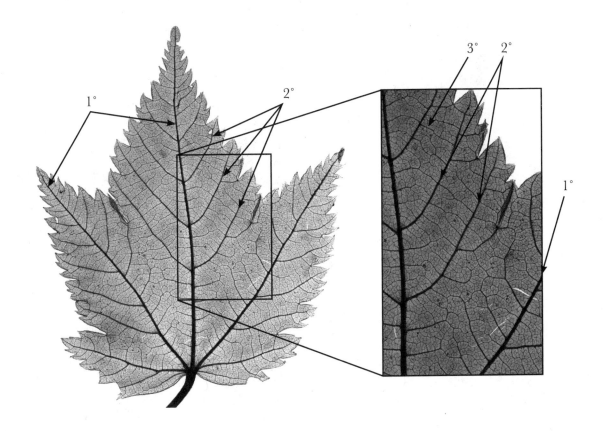

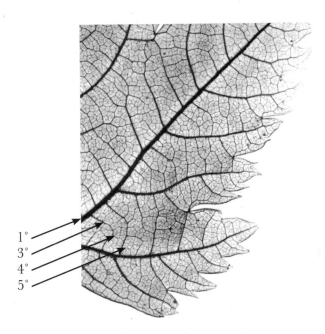

Fig. 140
Acer argutum
(Sapindaceae)

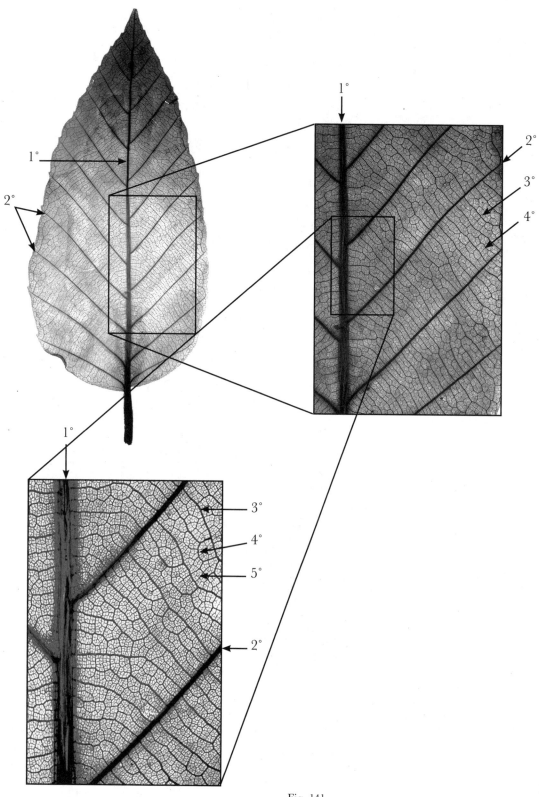

Fig. 141
Fagus longipetiolata
(Fagaceae)

General rules

Most leaves have a continuous sequence of vein orders that are typically easy to recognize by starting at the thickest (1° vein) and progressing to the finest. To recognize the 1°, 2°, and 3° veins, take the following steps

1. Find the vein or veins of the largest gauge: the *primary vein(s)* (some leaves have more than one). Most leaves have a single primary vein that gives rise to pinnately arranged *secondaries* or *costal veins* (in this case, go to step 3). If more than one vein originates at or near the base of the leaf, follow step 2 to determine if the leaf has more than one 1° vein.

2. After recognizing the single vein of greatest gauge as the primary (generally the midvein), other primaries are recognized by being at least 75% of the gauge of the widest primary (at the point of divergence from the widest primary). These veins are basal or nearly basal. If these veins enter lateral lobes or run in strong arches toward the apex, they are generally easily recognized as primaries. But if the lateral primaries curve toward the midvein distally (Figs. 142, 143) or branch toward the margin (Fig. 144), it may be hard to designate them as primaries or secondaries.

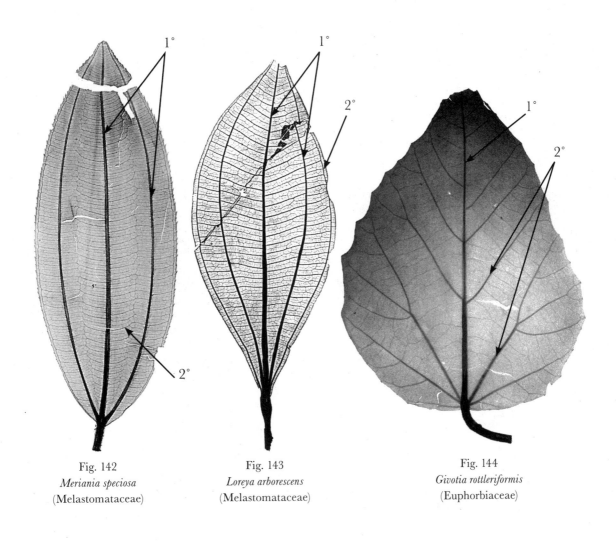

Fig. 142
Meriania speciosa
(Melastomataceae)

Fig. 143
Loreya arborescens
(Melastomataceae)

Fig. 144
Givotia rottleriformis
(Euphorbiaceae)

Note: If there is more than one 1° vein (based on vein gauge), other veins originating at the base may be considered primaries if their course is similar to that of the previously defined primaries, even if their gauge falls into the range of 25–75% of the widest 1° vein. If these veins are narrower than 25% of the widest 1° vein, they are not considered primaries (Figs. 145, 146).

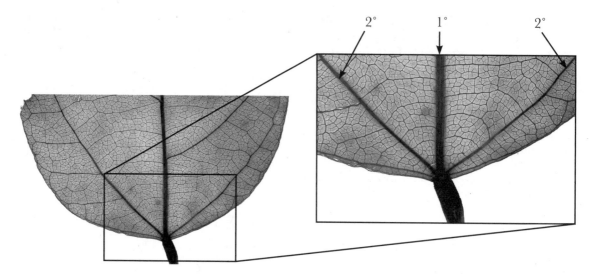

Fig.145
Pinnate venation
Tannodia swynnertonii
(Euphorbiaceae)

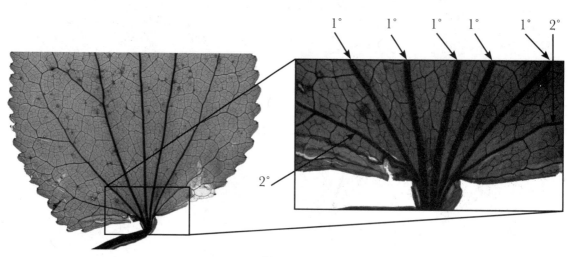

Fig. 146
Palmate venation with five 1° veins
Tetracentron sinense
(Trochodendraceae)

3. Find the veins of greatest gauge that form the vein field mesh or fabric of the leaf: the *tertiary veins* (Figs. 140, 141). Note that in some instances 2° veins fill the field (Figs. 142, 147). Tertiary veins are considered:

- **epimedial** if they intersect a 1° vein (Fig. 148).

- **intercostal** if they intersect a 2° vein but no primary (Fig. 148).

- **exterior** if they are exmedial to all 2° veins (Fig. 148).

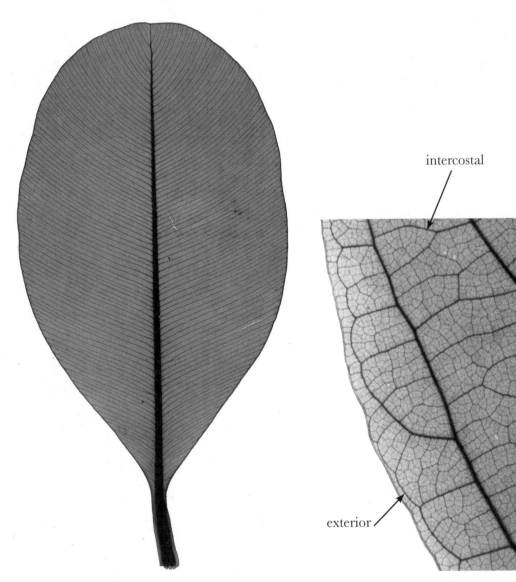

intercostal   epimedial

exterior

Fig. 147
Field filled by secondaries rather than tertiaries
Calophyllum calaba
(Clusiaceae)

Fig. 148
Tertiary veins
Sassafras albidum
(Lauraceae)

4. Having recognized the limits of the 1° and 3° vein sets, identify the set of veins that is intermediate in gauge. These are the *secondary veins,* and they may vary substantially in both gauge and course. Typical types of secondary veins include the following:

- **major (or costal) secondaries,** the rib-forming veins that originate on the primary and run toward the margins (Fig. 149).

- **minor secondaries,** which branch from lateral primaries or major secondaries and run toward the margins (Fig. 149). Note: these are often the "tines" of agrophic veins (see II.26).

- **interior secondaries,** which run between primaries in palmately veined leaves (Figs. 150, 151).

- **intersecondaries,** which have courses similar to major secondaries but have a gauge intermediate between secondaries and tertiaries and do not reach the margin (Fig. 152).

- **intramarginal secondaries,** which run parallel to the leaf margin with laminar tissue exmedial to them (Figs. 153).

- **marginal secondaries,** veins of secondary gauge that run on the margin of the leaf with no exmedial laminar tissue (Fig. 154; marginal veins of tertiary gauge are called *fimbrial veins,* see II.30.3).

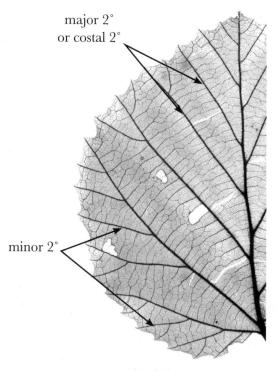

major 2°
or costal 2°

minor 2°

Fig. 149
Major and minor 2° veins
Parrotia jacquemontiana
(Hamamelidaceae)

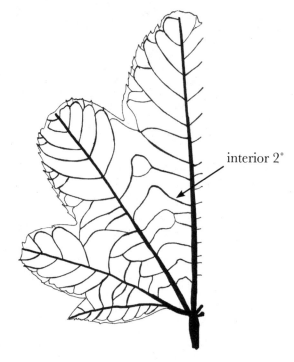

interior 2°

Fig. 150
Interior 2° veins

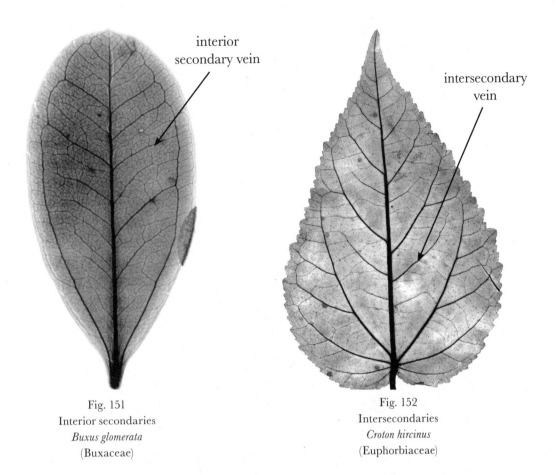

interior
secondary vein

intersecondary
vein

Fig. 151
Interior secondaries
Buxus glomerata
(Buxaceae)

Fig. 152
Intersecondaries
Croton hircinus
(Euphorbiaceae)

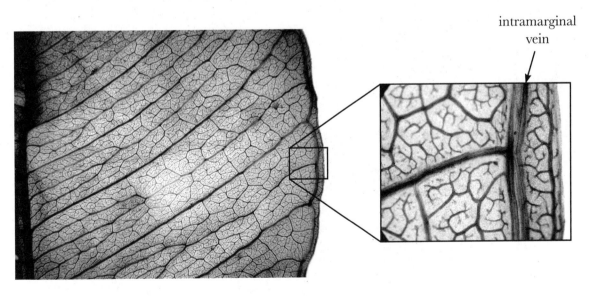

intramarginal
vein

Fig. 153
Intramarginal secondary
Spondias globosa
(Anacardiaceae)

5. Once you have recognized the first three orders of venation, proceed in sequence to determine the higher orders of venation using the criteria of vein gauge and course. Each successive vein order should have a distinctly narrower gauge, and the course may differ as well.

6. In most leaves, the veins of the finest gauge are *freely ending veinlets* (FEVs). FEVs enter, but do not cross, the smallest vein-bounded regions of leaf tissue, the *areoles*. FEVs can be unbranched, but they most often ramify within the areole. The boundaries of most areoles are formed by the highest order of excurrently branched veins.

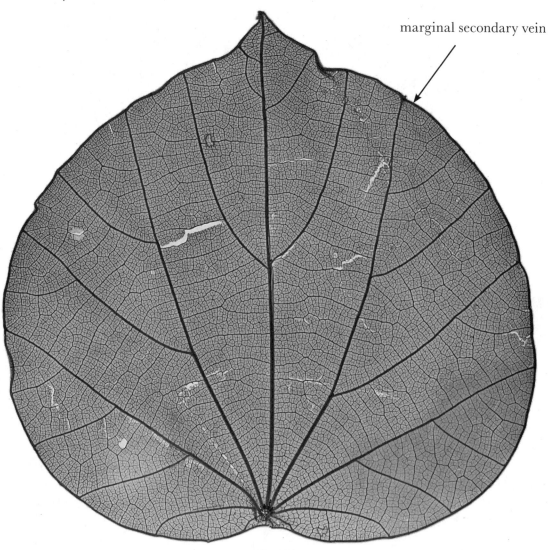

marginal secondary vein

Fig. 154
Marginal secondary
Diploclisia kunstleri
(Menispermaceae)

II. Vein Characters

23.   Primary Vein Framework

23.1   Pinnate – Leaf or leaflet has a single 1° vein (Figs. 155–158).

23.2   Palmate – Leaf has three or more basal veins, of which at least two are primaries (i.e., at least one has 75% of the gauge of the thickest vein, which is usually the midvein, see Determining Vein Order and Type, above). It can be difficult to distinguish pinnate from palmate primary frameworks near the 75% cutoff.

23.2.1   Actinodromous – Three or more 1° veins diverge radially from a single point.

23.2.1.1 Basal – Primary veins radiate from the petiolar insertion point (Figs. 159, 160).

23.2.1.2 Suprabasal – Primary veins radiate from a point distal to petiolar insertion (Fig. 161).

23.2.2   Palinactinodromous – Three or more primaries diverge in a series of branches rather than from a single point (Figs. 162, 163).

23.2.3   Acrodromous – Three or more primaries originate from a point and run in convergent arches toward the leaf apex.

23.2.3.1 Basal – Primary veins radiate from the petiolar insertion point (Figs. 164, 165).

23.2.3.2 Suprabasal – Primary veins radiate from a point distal to petiolar insertion (Fig. 166).

23.2.4   Flabellate – Several to many equally fine basal veins diverge radially at low angles to each other and branch distally (Fig. 167).

23.2.5   Parallelodromous (typically only in monocot leaves) – Multiple parallel 1° veins originate collaterally at the leaf base and converge toward the leaf apex (Fig. 168).

23.2.6   Campylodromous (typically only in monocot leaves) – Multiple parallel 1° veins originate collaterally at or near the leaf base and run in strongly recurved arches that converge toward the leaf apex (Fig. 169).

Fig. 155
Pinnate
Ostrya guatemalensis
(Betulaceae)

Fig. 156
Pinnate
Carrierea calycina
(Salicaceae)

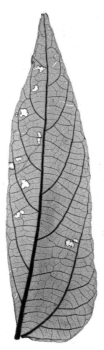

Fig. 157
Pinnate
Dalechampia cissifolia
(Euphorbiaceae)

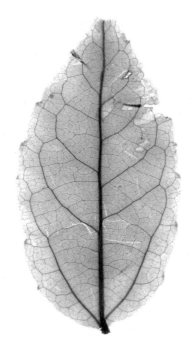

Fig. 158
Pinnate
Croton hircinus
(Euphorbiaceae)

Fig. 159
Basal actinodromous
Dombeya elegans
(Malvaceae)

Fig. 160
Basal actinodromous
Tetrameles nudiflora
(Datiscaceae)

Fig. 161
Suprabasal actinodromous
Phoebe costaricana
(Lauraceae)

Fig. 162
Palinactinodromous
Platanus racemosa
(Platanaceae)

Fig. 163
Palinactinodromous
Trichosanthes formosana
(Curcurbitaceae)

Fig. 164
Basal acrodromous
Paliurus ramosissimus
(Rhamnaceae)

Fig. 165
Basal acrodromous
Sarcorhachis naranjoana
(Piperaceae)

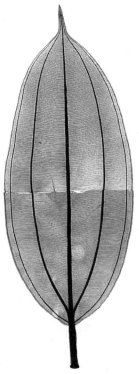

Fig. 166
Suprabasal acrodromous
Topobea watsonii
(Melastomataceae)

Fig. 167
Flabellate
Paranomus sceptrum
(Proteaceae)

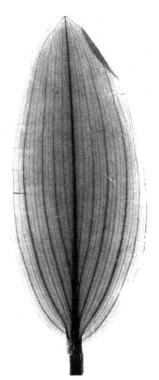

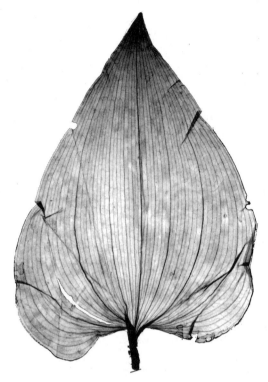

Fig. 168
Parallelodromous
Potamogeton amplifolius
(Potamogetonaceae)

Fig. 169
Campylodromous
Maianthemum dilatatum
(Ruscaceae)

24.   Naked Basal Veins

24.1   Absent (Figs. 165, 166).

24.2   Present – the exmedial side of one or both lateral primaries, or of basal secondaries, forms part of the leaf margin at the base (Fig. 170).

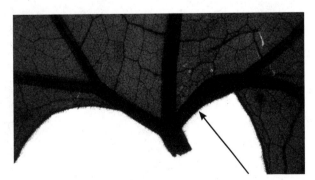

Fig. 170
Naked basal primary veins
Trichosanthes formosana
(Curcurbitaceae)

25.   **Number of Basal Veins** – Total number of 1° and 2° veins that originate in the base of the leaf and have courses similar to the course(s) of the primary or primaries. The leaf in figure 171 has six basal veins; the leaf in figure 172 has one basal vein.

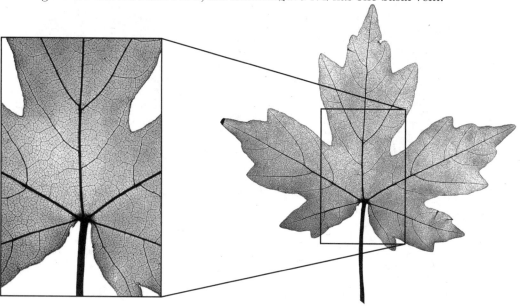

Fig. 171
Six basal veins
Acer miyabei
(Sapindaceae)

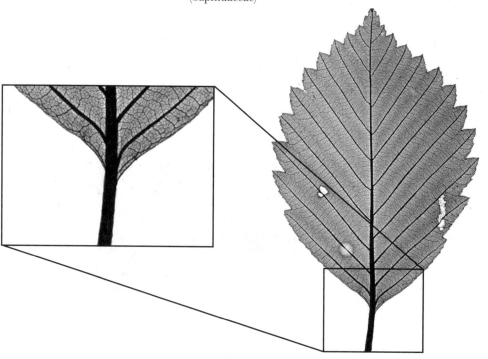

Fig. 172
One basal vein
Sorbus japonica
(Rosaceae)

26.   Agrophic Veins – A comblike complex of veins composed of a lateral 1° or 2° vein with two or more excurrent minor 2° veins that originate on it and travel roughly parallel courses toward the margin. The latter may be straight or looped and are only exterior (not bilaterally paired along the vein of origin). Agrophic veins are similar to pectinal veins as defined by Spicer (1986).

26.1   Absent (Figs. 172, 173)

26.2   Present

26.2.1 Simple – One or two agrophic veins. These may be paired (Fig. 174) or appear on only one half of the leaf.

26.2.2 Compound – More than two agrophic veins (Figs. 175, 176).

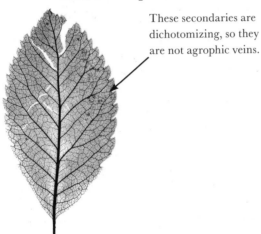

These secondaries are dichotomizing, so they are not agrophic veins.

Fig. 173
Agrophic veins absent
Eucryphia glandulosa
(Cunoniaceae)

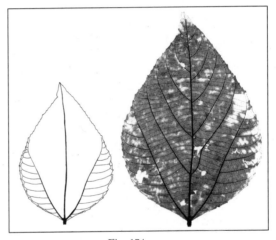

Fig. 174
Simple agrophic veins
Alchornea tiliifolia
(Euphorbiaceae)

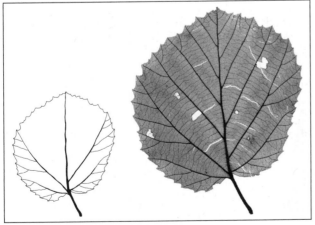

Fig. 175
Compound agrophic veins
Parrotia jacquemontiana
(Hamamelidaceae)

Fig. 176
Compound agrophic veins
Cissus caesia
(Vitaceae)

27.   **Major Secondary Vein Framework** – To describe 2° vein framework characters, examine the courses of the 2° veins in the middle of the lamina. There is no obligate relationship between secondary course and margin type: all major types of secondary course occur in both entire-margined and toothed leaves. When secondaries branch dichotomously, the branches are also considered to be secondaries. This is important in distinguishing eucamptodromous from semicraspedodromous secondaries, for example.

27.1   **Major secondaries (or their branches) reach the margin.**

27.1.1   **Craspedodromous** – Secondaries terminate at the margin (Figs. 177, 178) or at the marginal vein. It is possible, although rare, to have both craspedodromous secondaries and an entire margin (Fig. 179).

27.1.2   **Semicraspedodromous (usually in toothed leaves)** – Secondaries branch near the margin; one of the branches terminates at the margin, and the other joins the superjacent secondary (Figs. 180–182).

27.1.3   **Festooned semicraspedodromous** – Secondaries form more than one set of loops, with branches from the most exmedial loops terminating at the margin (Figs. 183–185).

27.2   **Major secondaries and their branches do not reach the margin and lose gauge by attenuation.**

27.2.1   **Eucamptodromous** – Secondaries connect to superjacent major secondaries via tertiaries without forming marginal loops of secondary gauge (Figs. 186–188). Three special cases are noted.

27.2.1.1 **Basal eucamptodromous** – All eucamptodromous secondaries arise from the base of the leaf ($<0.25L$; Fig. 189). May be difficult to distinguish from acrodromous primaries (II.23.2.3; Figs. 164–166).

27.2.1.2 **Hemieucamptodromous** – All eucamptodromous secondaries arise from the proximal half of the leaf (Fig. 190).

27.2.1.3 **Eucamptodromous becoming brochidodromous distally** – Proximal secondaries are eucamptodromous, but distal secondaries form loops of secondary gauge (Fig. 191).

27.2.2   **Reticulodromous** – Secondaries branch into a reticulum of higher-order veins (Fig. 192).

27.2.3   **Cladodromous** – Secondaries freely ramify exmedially (Fig. 193).

27.3   Major secondaries form loops of secondary gauge and do not reach the margin.

27.3.1 **Simple brochidodromous** – Secondaries join in a series of prominent arches or loops of secondary gauge. Junctions between secondaries are excurrent and the smaller vein has >25% of the gauge of the larger vein at the junction (Figs. 194–196).

27.3.2 **Festooned brochidodromous** – Secondaries branch into multiple sets of loops of secondary gauge, often with accessory loops of higher gauge (Figs. 197–199).

27.4   Mixed – Major secondary course varies within the leaf (Fig. 200).

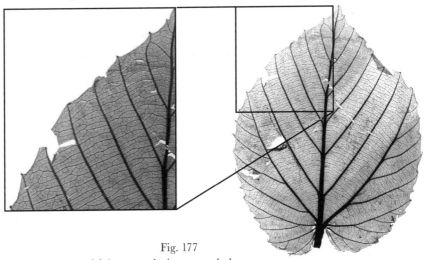

Fig. 177
Major secondaries craspedodromous
Corylopsis glabrescens
(Hamamelidaceae)

Fig. 178
Major secondaries craspedodromous
Desfontainea spinosa
(Desfontaineaceae)

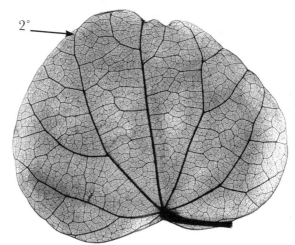

Fig. 179
Major secondaries craspedodromous
Cyclea merrillii
(Menispermaceae)

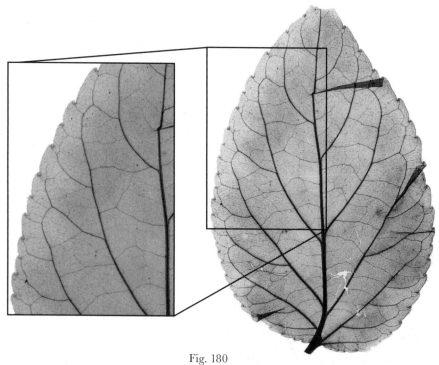

Fig. 180
Major secondaries semicraspedodromous
Aphaerema spicata
(Salicaceae)

Fig. 181
Semicraspedodromous
Cercidiphyllum japonicum
(Cercidiphyllaceae)

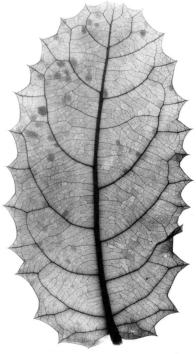

Fig. 182
Semicraspedodromous
Casearia ilicifolia
(Salicaceae)

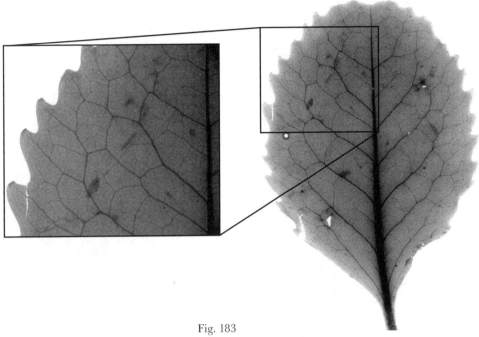

Fig. 183
Major secondaries festooned semicraspedodromous
Laurelia novae-zelandiae
(Atherospermataceae)

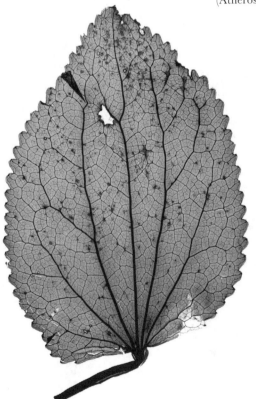

Fig. 184
Festooned semicraspedodromous
Tetracentron sinense
(Trochodendraceae)

Fig. 185
Festooned semicraspedodromous
Mahonia wilcoxii
(Berberidaceae)

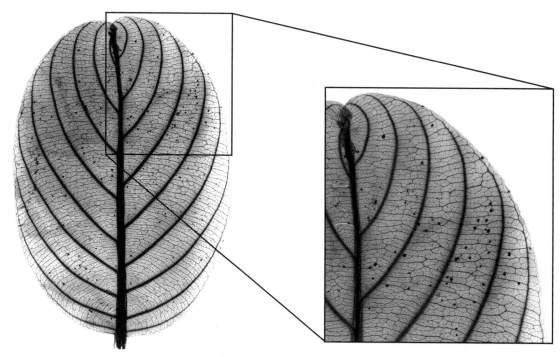

Fig. 186
Major secondaries eucamptodromous
Tetracera rotundifolia
(Dilleniaceae)

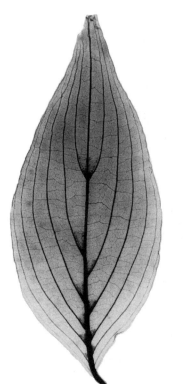

Fig. 187
Eucamptodromous
Cornus officinalis
(Cornaceae)

Fig. 188
Eucamptodromous
Isoptera lissophylla
(Dipterocarpaceae)

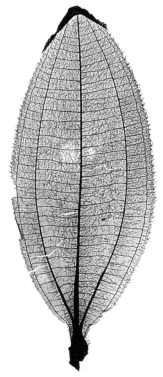

Fig. 189
Basal eucamptodromous
Tococa aristata
(Melastomataceae)

brochidodromous

eucamptodromous

Fig. 190
Hemieucamptodromous
Cleistanthus oligophlebius
(Phyllanthaceae)

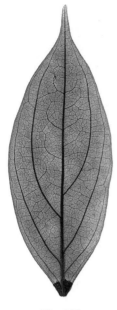

Fig. 191
Eucamptodromous becoming brochidodromous distally
Rhamnidium elaeocarpum
(Rhamnaceae)

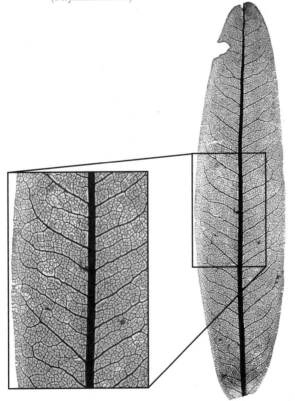

Fig. 192
Reticulodromous
Eucryphia moorei
(Cunoniaceae)

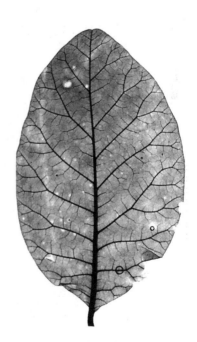

Fig. 193
Cladodromous
Cotinus obovatus
(Anacardiaceae)

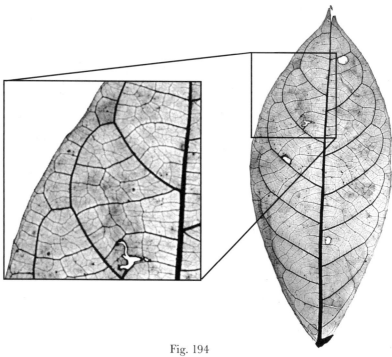

Fig. 194
Major secondaries brochidodromous
Baccaurea staudtii
(Phyllanthaceae)

Fig. 195
Brochidodromous
Santiria samarensis
(Burseraceae)

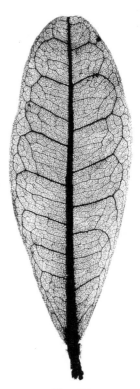

Fig. 196
Brochidodromous
Aextoxicon punctatum
(Aextoxicaceae)

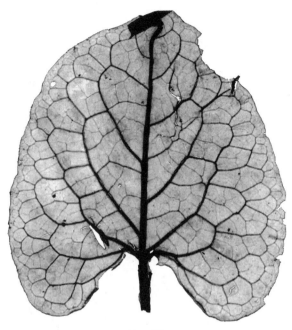

Fig. 197
Major secondaries festooned brochidodromous
Antigonon cinerascens
(Polygonaceae)

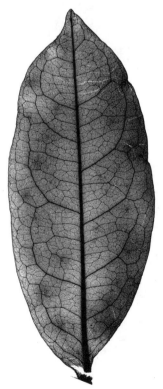

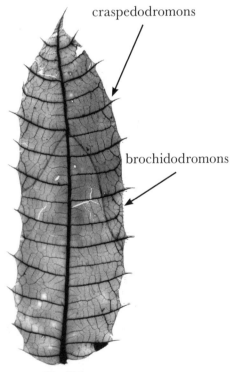

craspedodromons

brochidodromons

Fig. 198
Major secondaries festooned
brochidodromous
Capsicodendron pimenteira
(Canellaceae)

Fig. 199
Major secondaries festooned
brochidodromous
Tapura guianensis
(Dichapetalaceae)

Fig. 200
Major secondaries mixed
Comocladia glabra
(Anacardiaceae)

28.   Interior Secondaries

28.1   Absent (Figs. 181, 197, 205).

28.2   Present – These secondaries cross between 1° veins or between 1° and perimarginal 2° veins (see II.30) but do not reach the margin (Figs. 201–203). They are typically arched or straight and are present in the central part of many palmately lobed leaves, where they may have a course similar to adjacent 3° veins. Interior secondaries may also occur in leaves with acrodromous 1° veins, intramarginal 2° veins (Fig. 203), or basally eucamptodromous secondaries.

Fig. 201
Interior secondary
Filipendula occidentalis
(Rosaceae)

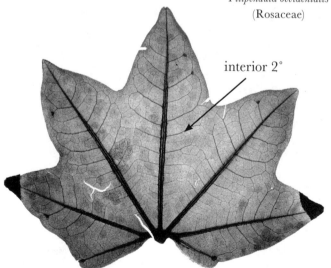

Fig. 202
Interior secondary
Triplochiton scleroxylon
(Malvaceae)

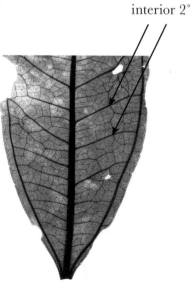

Fig. 203
Interior secondary
Scaphocalyx spathacea
(Achariaceae)

29.   **Minor Secondary Course**

29.1   **Craspedodromous** – Terminating at the margin (Fig. 204).

29.2   **Simple brochidodromous** – Joined together in a series of prominent arches or loops of secondary gauge. Junctions between secondaries are excurrent and the smaller vein has >25% of the gauge of the larger (Fig. 205).

29.3   **Semicraspedodromou**s – Minor secondaries branch near the margin. One of the branches eventually terminates at the margin, and the other joins the superjacent minor secondary (Fig. 206).

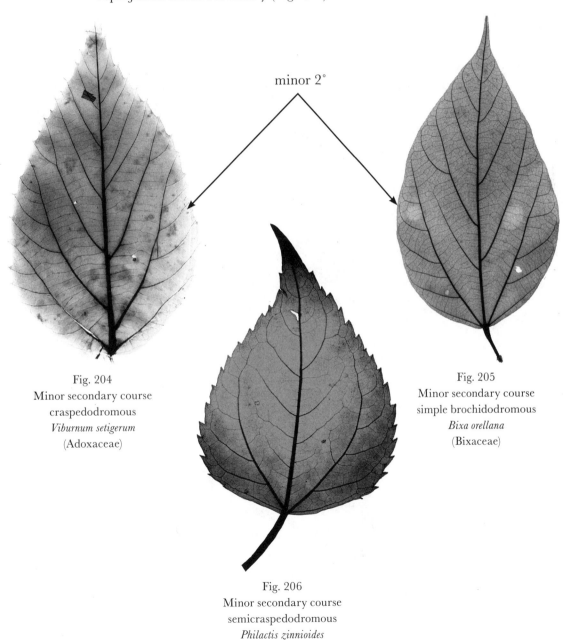

minor 2°

Fig. 204
Minor secondary course
craspedodromous
Viburnum setigerum
(Adoxaceae)

Fig. 205
Minor secondary course
simple brochidodromous
Bixa orellana
(Bixaceae)

Fig. 206
Minor secondary course
semicraspedodromous
Philactis zinnioides
(Asteraceae)

30.   **Perimarginal Veins** – When present, these veins closely parallel the leaf margin and lose little gauge distally.

   30.1   **Marginal secondary** – Vein of 2° gauge running on the leaf margin (Fig. 207). There are no veins exmedial to a marginal secondary.

   30.2   **Intramarginal secondary** – Vein of 2° gauge running near the margin with laminar tissue exmedial to it (Figs. 208, 209). Intramarginal veins typically are intersected by major secondaries.

   30.3   **Fimbrial vein** – Vein of consistent 3° gauge running on the margin with no laminar tissue exmedial to it (Fig. 210).

marginal 2°   intramarginal 2°   intramarginal 2°

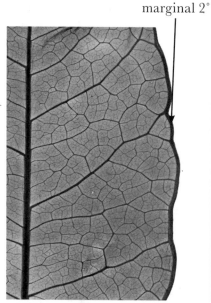 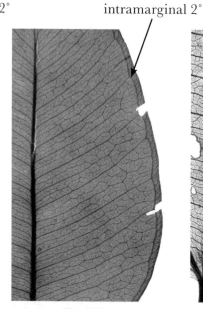 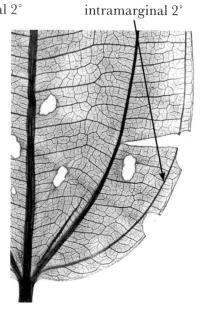

Fig. 207
Marginal secondary
Securidaca marginata
(Polygalaceae)

Fig. 208
Intramarginal secondary
Spondias bivenomarginalis
(Anacardiaceae)

Fig. 209
Intramarginal secondary
Graffenrieda anomala
(Melastomataceae)

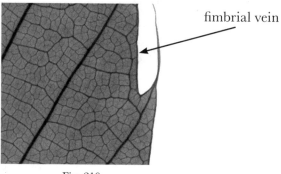

fimbrial vein

Fig. 210
Fimbrial vein
Castanea sativa
(Fagaceae)

31.   **Major Secondary Spacing** – Variation in the distance between adjacent secondaries, measured at their intersections with the midvein.

> **31.1**   **Regular** – Secondary spacing proportionally decreases distally and proximally (Fig. 211).
>
> **31.2**   **Irregular** – Secondary spacing varies over the lamina (Fig. 212).
>
> **31.3**   **Decreasing proximally** – Secondary spacing decreases toward base (Fig. 213); may be regular or irregular.
>
> **31.4**   **Gradually increasing proximally** – Secondary spacing increases gradually toward base (Fig. 214).
>
> **31.5**   **Abruptly increasing proximally** – Secondary spacing increases abruptly toward base (Fig. 215).

Fig. 211
Secondary spacing regular
Vitex limonifolia
(Lamiaceae)

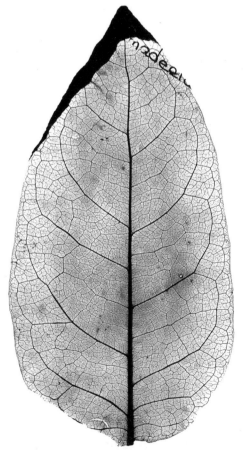

Fig. 212
Secondary spacing irregular
Kermadecia sinuata
(Proteaceae)

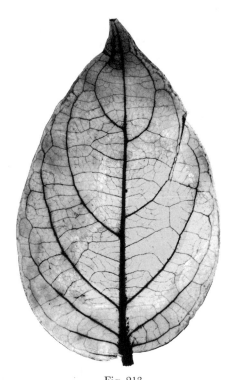

Fig. 213
Secondary spacing decreasing proximally
Glochidion bracteatum
(Phyllanthaceae)

Fig. 214
Secondary spacing gradually increasing proximally
Populus jackii
(Salicaceae)

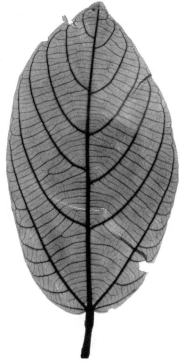

Fig. 215
Secondary spacing abruptly increasing proximally
Apeiba macropetala
(Malvaceae)

32.   Variation of Major Secondary Angle to Midvein – Each angle measured on the distal side of the junction (the vertex) of the secondary with the midvein. One ray of the angle follows the midvein distal to the junctions, and the other follows the secondary for 25% of its length. The major secondary angle should be evaluated proximal to 0.75 l_m:

32.1   Uniform – Major 2° angle varies <10° from the base to 0.75 l_m (Fig. 216).

32.2   Inconsistent – Major 2° angle varies >10° from the base to 0.75 l_m (Fig. 217).

32.3   Smoothly increasing proximally (Fig. 218).

32.4   Smoothly decreasing proximally (Fig. 219).

32.5   Abruptly increasing proximally (Fig. 220).

32.6   One pair of acute basal secondaries (Figs. 215, 221).

Fig. 216
Secondary angle uniform
Tilia heterophylla
(Malvaceae)

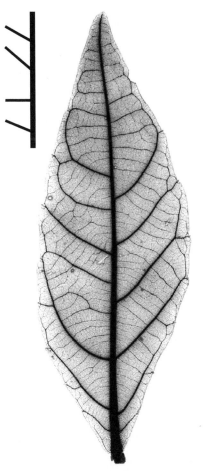

Fig. 217
Secondary angle inconsistent
Alchornea polyantha
(Euphorbiaceae)

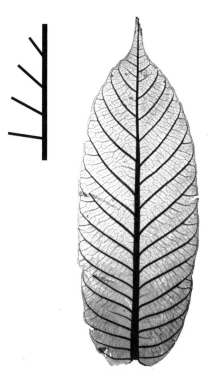

Fig. 218
Secondary angle smoothly increasing proximally
Pseudolmedia laevis
(Moraceae)

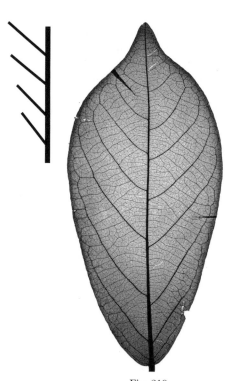

Fig. 219
Secondary angle smoothly decreasing proximally
Popowia congensis
(Annonaceae)

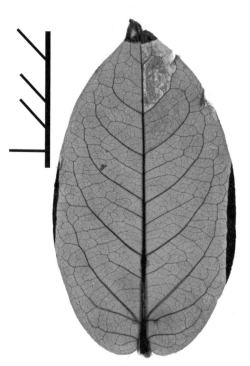

Fig. 220
Secondary angle abruptly increasing proximally
Banisteriopsis laevifolia
(Malpighiaceae)

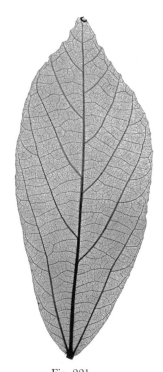

Fig. 221
One pair of acute basal secondaries
Microcos tomentosa
(Malvaceae)

33.   Major Secondary Attachment to Midvein

33.1   Decurrent – Major secondaries meet the midvein asymptotically (Fig. 129, 222).

33.2   Proximal secondaries decurrent – Major secondaries near the lamina base are decurrent on midvein, though distal secondaries are excurrent (Fig. 223).

33.3   Excurrent – Major secondaries join the midvein without deflecting it, midvein monopodial (Fig. 224).

33.4   Deflected – Midvein is deflected at junctions with major secondaries and is thus sympodial (Fig. 225).

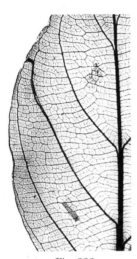

Fig. 222
Decurrent secondary attachment
Itea chinensis
(Iteaceae)

Fig. 223
Proximal secondaries decurrent
Crataegus brainerdii
(Rosaceae)

Fig. 224
Excurrent secondary attachment
Tetracera podotricha
(Dilleniaceae)

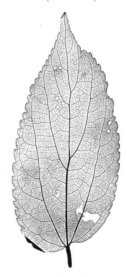

Fig. 225
Deflected secondary attachment
Celtis cerasifera
(Cannabaceae)

34.   **Intersecondary Veins** – Veins with courses similar to those of the major secondaries, but generally shorter in exmedial extent and intermediate in gauge between major secondaries and tertiaries (Fig. 226).

34.1   **Intersecondary proximal course**

34.1.1   **Parallel to major secondarie**s (Fig. 227).

34.1.2   **Perpendicular to midvein** (Fig. 228).

34.2   **Intersecondary length**

34.2.1   **Less than 50% of subjacent secondary** (Fig. 229).

34.2.2   **More than 50% of subjacent secondary** (Fig. 230).

34.3   **Intersecondary distal course**

34.3.1   **Reticulating or ramifying** – Branching and losing a defined course (Fig. 231).

34.3.2   **Parallel to a major secondary** (Fig. 232).

34.3.3   **Perpendicular to a subjacent major secondary** (Fig. 233).

34.3.4   **Basiflexed but not joining the subjacent secondary at right angles** (Fig. 234).

34.4   **Intersecondary frequency** – Average number of intersecondary veins per intercostal area

34.4.1   **Less than one per intercostal area** (Fig. 235).

34.4.2   **Usually one per intercostal area** (Fig. 236).

34.4.3   **More than one per intercostal area** (Fig. 237).

intersecondary

Fig. 226
Intersecondary veins
Couepia paraensis
(Chrysobalanaceae)

intersecondary

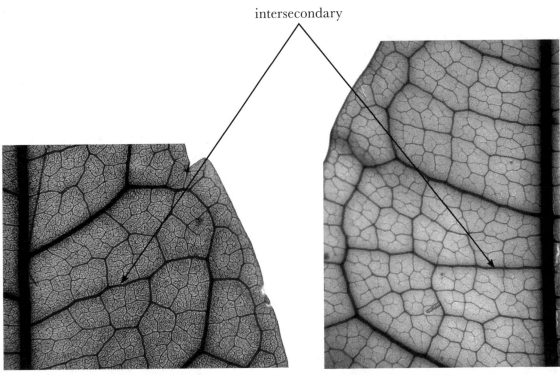

Fig. 227
Proximal course of intersecondary
parallel to major secondaries
Protium subserratum
(Burseraceae)

Fig. 228
Proximal course of intersecondary
perpendicular to midvein
Dacryodes negrensis
(Burseraceae)

intersecondary

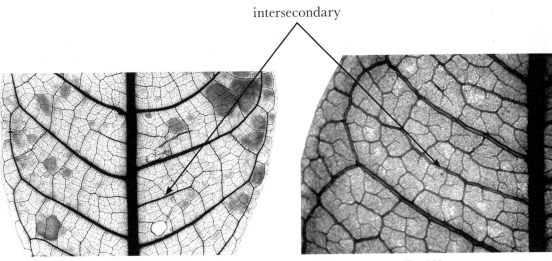

Fig. 229
Length of intersecondary <50% of subjacent secondary
Protium opacum
(Burseraceae)

Fig. 230
Length of intersecondary >50% of subjacent secondary
Santiria griffithii
(Burseraceae)

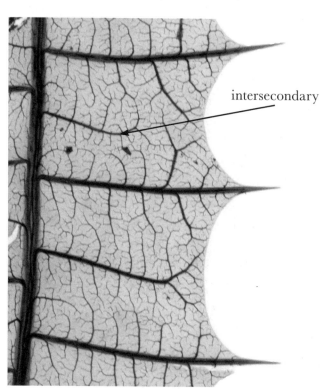

intersecondary

Fig. 231
Distal course of intersecondary
reticulating or ramifying
Comocladia cuneata
(Anacardiaceae)

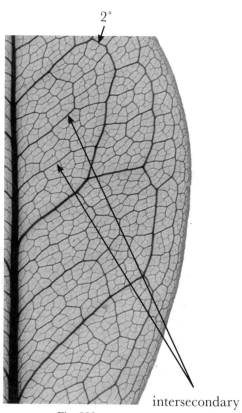

2°

intersecondary

Fig. 232
Distal course of intersecondary
parallel to major secondary
Ancistrocladus tectorius
(Ancistrocladaceae)

intersecondary

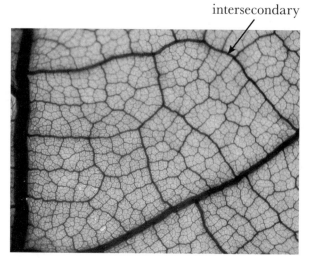

Fig. 233
Distal course of intersecondary perpendicular
to subjacent major secondary
Canarium ovatum
(Burseraceae)

intersecondary

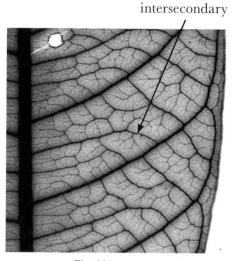

Fig. 234
Distal course of intersecondary basiflexed
Stemonoporus nitidus
(Dipterocarpaceae)

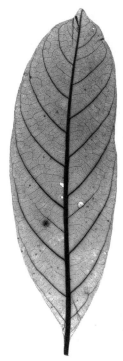

Fig. 235
Frequency of intersecondary veins
<1 per intercostal area
Guarea tuberculata
(Meliaceae)

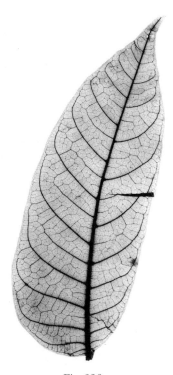

Fig. 236
Frequency of intersecondary veins
~1 per intercostal area
Cedrela angustifolia
(Meliaceae)

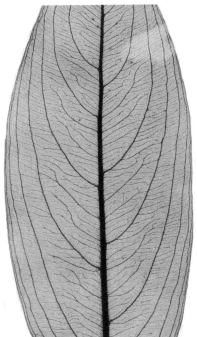

Fig. 237
Frequency of intersecondary veins >1 per intercostal area
Ouratea aff. *garcinioides*
(Ochnaceae)

35.   **Intercostal Tertiary Vein Fabric** – The three major categories are percurrent (35.1), reticulate (35.2), and ramified (35.3).

35.1   **Percurrent** – Tertiaries cross between adjacent secondaries.

35.1.1   **Course of percurrent tertiaries**

35.1.1.1 **Opposite** – Majority of tertiaries cross between adjacent secondaries in parallel paths without branching (Figs. 238–241).

35.1.1.1.1 **Straight** – Passing across the intercostal area without a noticeable change in course (Fig. 238).

35.1.1.1.2 **Convex** – Middle portion of the vein arches exmedially, without an inflection point (Fig. 239).

35.1.1.1.3 **Sinuous** – Changes direction of curvature (Fig. 240).

35.1.1.1.4 **Forming a chevron** – Most tertiary courses have a markedly sharp bend (Fig. 241).

35.1.1.2 **Alternate** – Majority of tertiaries cross between secondaries with regular offsets (abrupt angular discontinuities) near the middle of the intercostal area (Fig. 242).

35.1.1.3 **Mixed** – Tertiaries have both opposite and alternate percurrent courses (Fig. 243).

35.1.2   **Angle of percurrent tertiaries** – Angle formed between the midvein trend and the course of a percurrent 3° vein projected to the midvein (Fig. 244).

35.1.2.1 **Acute** – Angle <90° (Fig. 245).

35.1.2.2 **Obtuse** – Angle >90° (Fig. 246).

35.1.2.3 **Perpendicular** – Angle ~90° (Fig. 247).

35.2   **Reticulate** – Veins anastomose with other tertiary veins or secondary veins to form a net (Figs. 248, 249).

35.2.1   **Irregular** – Tertiaries anastomose at various angles to form irregular polygons (Fig. 248) or non-polygonal nets.

35.2.2   **Regular** – Tertiaries anastomose with other tertiaries at regular angles to generate a regular polygonal field (Fig. 249).

35.2.3   **Composite admedial** – Tertiaries connect to a trunk that ramifies admedially toward the axil of the subjacent costal secondary (Fig. 250).

35.3   **Ramified** – Tertiaries branch without forming a tertiary reticulum.

    35.3.1   Admedially ramified – Multiple tertiary veins branch toward the primary or midvein (Fig. 251).

    35.3.2   Exmedially ramified – Tertiary branching is oriented toward the leaf margin (Fig. 252).

    35.3.3   Transversly ramified – Opposed 3° veins from adjacent major secondaries ramify and join at a higher vein order (Fig. 253).

    35.3.4   Transversly freely ramified – Tertiary veins originating on one secondary vein branch toward adjacent secondary but do not join other veins from the opposing secondary (Fig. 254).

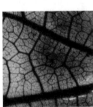

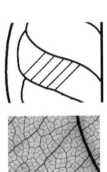
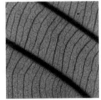
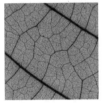

Fig. 238
Straight
Melicytus fasciger
(Violaceae)

Fig. 239
Convex
unknown genus
(Dipterocarpaceae)

Fig. 240
Sinuous
Aphelandra pulcherrima
(Acanthaceae)

Fig. 241
Forming a chevron
Vallea stipularis
(Elaeocarpaceae)

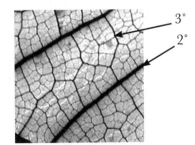

Fig. 242
Alternate percurrent tertiary fabric
Santiria samarensis
(Burseraceae)

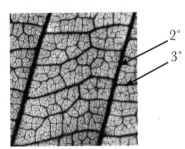

Fig. 243
Mixed percurrent tertiary fabric
Davilla rugosa
(Dilleniaceae)

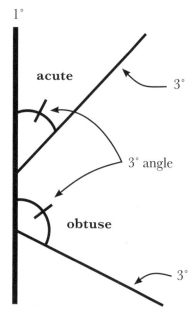

Fig. 244
Measurement of tertiary angle with
respect to the 1° vein

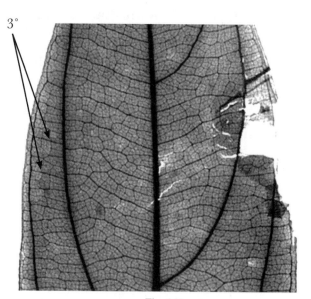

Fig. 245
Acute tertiary angles
Nectandra cuspidata
(Lauraceae)

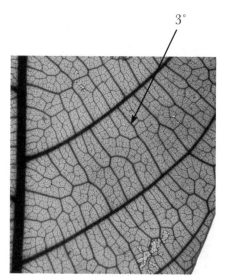

Fig. 246
Obtuse tertiary angle
Sloanea eichleri
(Elaeocarpaceae)

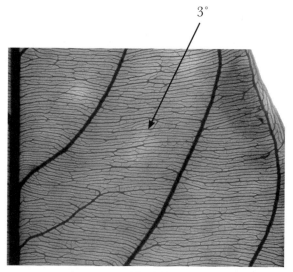

Fig. 247
Perpendicular tertiary angle
Bhesa archboldiana
(Celastraceae)

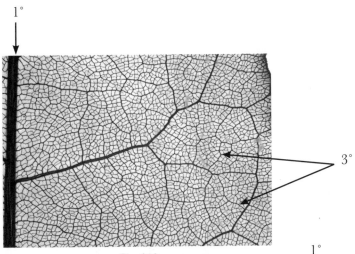

Fig. 248
Irregular reticulate tertiary fabric
Piranhea trifoliata
(Picrodendraceae)

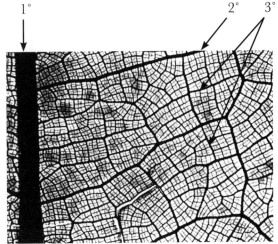

Fig. 249
Regular reticulate tertiary fabric
Afrostyrax kamerunensis
(Huaceae)

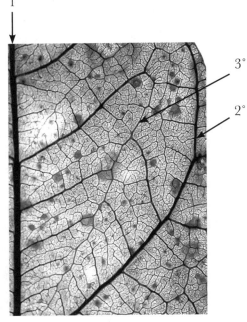

Fig. 250
Composite admedial
Sorindeia gilletii
(Anacardiaceae)

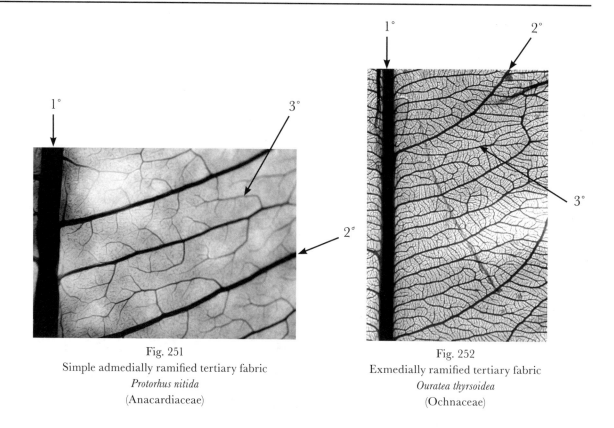

Fig. 251
Simple admedially ramified tertiary fabric
Protorhus nitida
(Anacardiaceae)

Fig. 252
Exmedially ramified tertiary fabric
Ouratea thyrsoidea
(Ochnaceae)

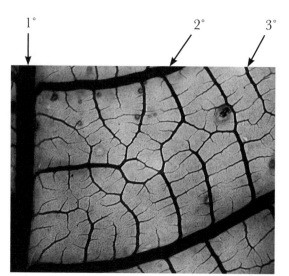

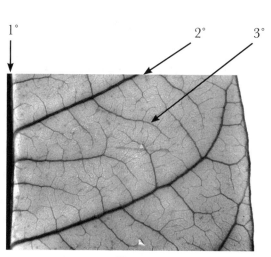

Fig. 253
Transversely ramified tertiary fabric
Comocladia glabra
(Anacardiaceae)

Fig. 254
Transversely freely ramified tertiary fabric
Rhus taitensis
(Anacardiaceae)

36.   **Intercostal Tertiary Vein Angle Variability** – Applies only to leaves with percurrent tertiaries; see 35.1.2 for measuring the angle. A leaf may exhibit more than one character state.

   36.1   **Inconsistent** – Angles of the tertiaries vary randomly over the lamina (Fig. 255).

   36.2   **Consistent** – Angles of the tertiaries do not vary over the surface of the lamina by more than 10% (Fig. 256).

   36.3   **Increasing exmedially** – Angles of the tertiaries become more obtuse away from the midvein (Fig. 257).

      36.3.1   **Basally concentric** – Special case of "increasing exmedially" such that the tertiaries form a "spider web pattern" around the primary vein(s) at the base of the leaf (Fig. 258).

   36.4   **Decreasing exmedially** – Angles of the tertiaries become more acute away from the midvein (Fig. 259).

   36.5   **Increasing proximally** – Angles of the tertiaries become more obtuse toward the base of the lamina (Fig. 260).

   36.6   **Decreasing proximally** – Angles of the tertiaries become more acute toward the base of the lamina (Fig. 261).

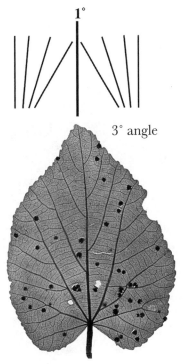

3° angle

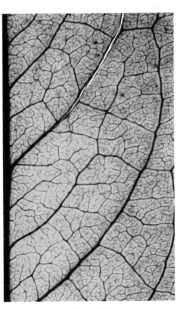

Fig. 255
Inconsistent tertiary angle
Viburnum sempervirens
(Adoxaceae)

Fig. 256
Consistent tertiary angle
Diospyros maritima
(Ebenaceae)

Fig. 257
Tertiary angle increasing
exmedially
Eriolaena malvacea
(Malvaceae)

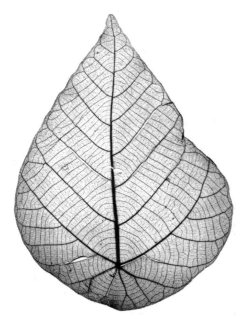

Fig. 258
Basally concentric
Macaranga bicolor
(Euphorbiaceae)

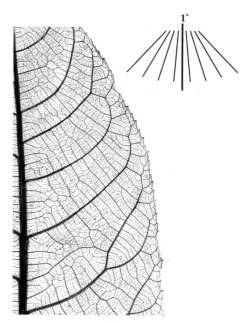

Fig. 259
Tertiary angle decreasing exmedially
Juglans boliviana
(Juglandaceae)

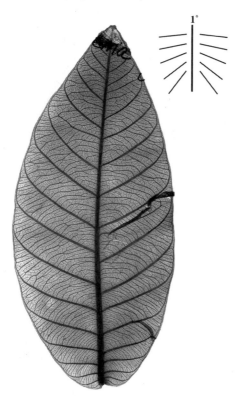

Fig. 260
Tertiary angle increasing proximally
Odontadenia geminata
(Apocynaceae)

Fig. 261
Tertiary angle decreasing proximally
Flacourtia rukam
(Salicaceae)

37.   **Epimedial Tertiaries** – Tertiaries that intersect a 1° vein.

37.1   **Epimedial tertiary fabric**

   37.1.1   Percurrent – Epimedial veins cross between 1° and 2° veins.

   37.1.1.1 Opposite percurrent – Majority of tertiaries cross between primary and secondaries in parallel paths without branching (Fig. 262).

   37.1.1.2 Alternate percurrent – Majority of tertiaries cross between primary and secondaries with regular offsets (abrupt angular discontinuities) (Fig. 263).

   37.1.1.3 Mixed – Approximately equal numbers of opposite and alternate percurrent tertiaries (Fig. 264).

   37.1.2   Ramified – Epimedial tertiaries branch toward the leaf margin (Fig. 265).

   37.1.3   Reticulate – Epimedial tertiaries anastomose with other 3° veins to form a net (Fig. 266).

   37.1.4   Mixed – Epimedial tertiaries do not consistently exhibit one characteristic (Fig. 267).

37.2   **Course of percurrent epimedial tertiaries**

   37.2.1   Proximal/admedial course of the epimedial tertiaries – Course of the epimedial tertiaries from their junction with the midvein to their approximate midpoint. More than one character state may apply.

   37.2.1.1 Parallel to the subjacent secondary (Fig. 268).

   37.2.1.2 Parallel to the intercostal tertiaries (Fig. 269).

   37.2.1.3 Perpendicular to the midvein (Fig. 270).

   37.2.1.4 Parallel to the intersecondary (Fig. 271).

   37.2.1.5 Obtuse to the midvein (Fig. 272).

   37.2.1.6 Acute to the midvein (Fig. 273).

   37.2.2   Distal/exmedial course of the epimedial tertiaries – Course of the epimedial tertiaries from their approximate midpoint to their intersection with the adjacent secondary (if not ramifying or reticulating). **Note:** More than one character state may apply.

   37.2.2.1 Parallel to intercostal tertiary – Epimedial tertiaries match pattern of adjacent intercostal tertiaries (Fig. 274).

37.2.2.2 Basiflexed – Course bends toward the base of the leaf and may either join the secondaries or lose gauge (Fig. 274, 275).

37.2.2.3 Acroflexed – Course bends toward the apex of the leaf and may either join the secondaries or lose gauge (Fig. 276).

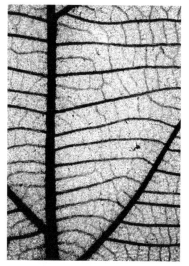

Fig. 262
Opposite percurrent
epimedial tertiaries
Actinidia latifolia
(Actinidiaceae)

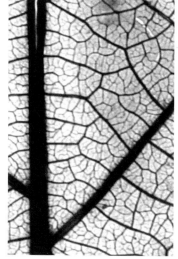

Fig. 263
Alternate percurrent
epimedial tertiaries
Alangium chinense
(Cornaceae)

Fig. 264
Mixed percurrent
epimedial tertiaries
Bixa orellana
(Bixaceae)

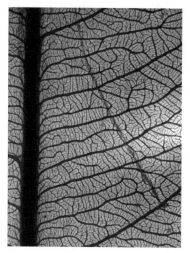

Fig. 265
Ramified epimedial tertiaries
Ouratea thyrsoidea
(Ochnaceae)

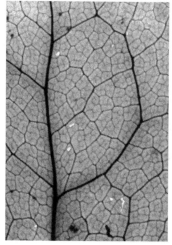

Fig. 266
Reticulate epimedial tertiaries
Mahonia wilcoxii
(Berberidaceae)

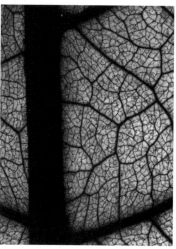

Fig. 267
Mixed epimedial tertiaries
Aphelandra pulcherrima
(Acanthaceae)

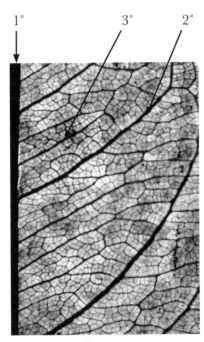

Fig. 268
Proximal course of the epimedial tertiaries
parallel to subjacent secondary
Capparis lundellii
(Capparaceae)

Fig. 269
Proximal course of the epimedial tertiaries
parallel to intercostal tertiary
Callicoma serratifolia
(Cunoniaceae)

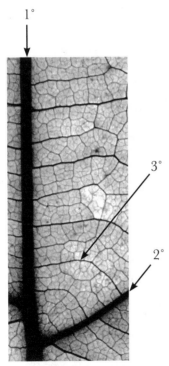

Fig. 270
Proximal course of the epimedial
tertiaries perpendicular to
the midvein
Lunania mexicana
(Salicaceae)

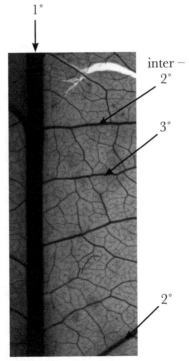

Fig. 271
Proximal course of the
epimedial tertiaries parallel to
the intersecondary
Celastrus racemosus
(Celastraceae)

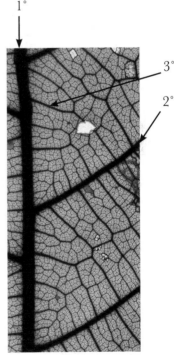

Fig. 272
Proximal course of the epimedial
tertiaries obtuse to the midvein
Sloanea eichleri
(Elaeocarpaceae)

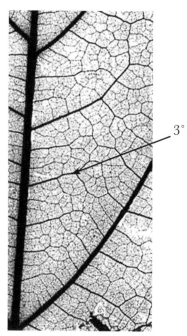

Fig. 273
Proximal course of the epimedial
tertiaries acute to midvein
Bixa orellana
(Bixaceae)

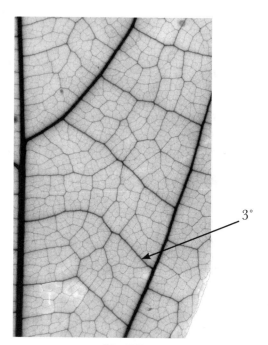

Fig. 274
Distal course of the epimedial tertiaries
parallel to intercostal tertiaries
Theobroma microcarpa
(Malvaceae)

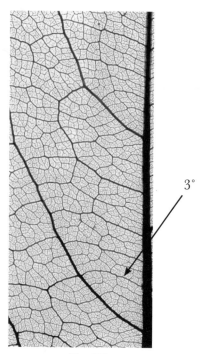

Fig. 275
Distal course of the epimedial tertiaries basiflexed
Spiropetalum erythrosepalum
(Connaraceae)

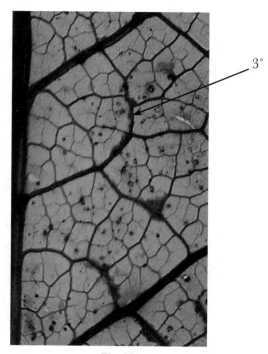

Fig. 276
Distal course of the epimedial tertiaries acroflexed
Commiphora aprevalii
(Burseraceae)

38. **Exterior Tertiary Course** – Configuration of the third-order veins that lie exmedially to the outermost secondaries but do not necessarily form the marginal ultimate veins.

   38.1   **Absent** – Leaf does not have exterior tertiaries (Fig. 277).

   38.2   **Looped** – Tertiaries form loops (Figs. 278, 279).

   38.3   **Terminating at the margin** – Tertiaries terminate at the margin (Figs. 280, 281).

   38.4   **Variable** – Pattern is not consistent (Fig. 282).

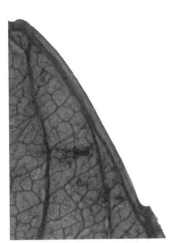

Fig. 277
Exterior tertiaries absent
Hedyosmum costaricense
(Chloranthaceae)

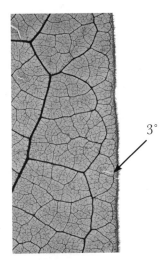

Fig. 278
Exterior tertiaries looped
Picramnia krukovii
(Picramniaceae)

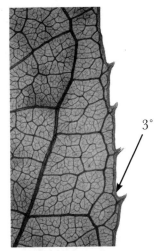

Fig. 279
Exterior tertiaries looped
Mollinedia floribunda
(Monimiaceae)

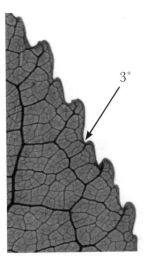

Fig. 280
Exterior tertiaries
terminating at margin
Barringtonia reticulata
(Lecythidaceae)

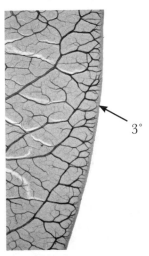

Fig. 281
Exterior tertiaries
terminating at margin
Carissa bispinosa
(Apocynaceae)

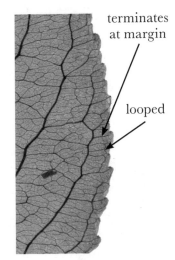

Fig. 282
Exterior tertiaries variable
Gymnosporia senegalensis
(Celastraceae)

39.   **Quaternary Vein Fabric** – Pattern formed by fourth-order vein courses. This and other higher-order venation characters should be scored near the center of the blade.

39.1   **Percurrent**

39.1.1   **Opposite** – Most quaternary veins cross between adjacent tertiary veins in parallel paths without branching (Fig. 283).

39.1.2   **Alternate** – Most quaternary veins cross between adjacent tertiaries with an offset (an abrupt angular discontinuity) (Fig. 284).

39.1.3   **Mixed percurrent** – Quaternaries are alternate and opposite in equal proportions (Fig. 285).

39.2   **Reticulate** – Quaternaries anastomose with other veins to form a net.

39.2.1   **Regular** – Angles formed by the vein intersections are regular (Fig. 286).

39.2.2   **Irregular** – Angles formed by the vein intersections are highly variable (Fig. 287).

39.3   **Freely ramifying** – Quaternaries branch freely and are the finest vein-order the leaf exhibits (Fig. 288).

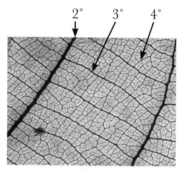

Fig. 283
Opposite percurrent quaternaries
Shorea congestiflora
(Dipterocarpaceae)

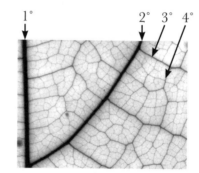

Fig. 284
Alternate percurrent quaternaries
Theobroma microcarpa
(Malvaceae)

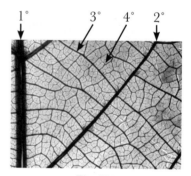

Fig. 285
Mixed percurrent quaternary
Alangium chinense
(Cornaceae)

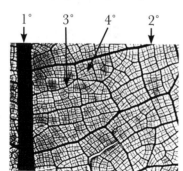

Fig. 286
Regular reticulate quaternaries
Afrostyrax kamerunensis
(Huaceae)

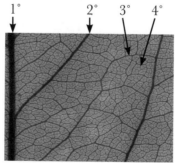

Fig. 287
Irregular reticulate quaternaries
Diospyros pellucida
(Ebenaceae)

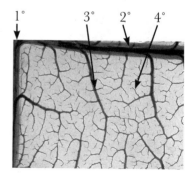

Fig. 288
Freely ramifying quaternary
Comocladia cuneata
(Anacardiaceae)

40. **Quinternary Vein Fabric** – Pattern formed by 5° vein courses, when present. This and other higher-order venation characters should be scored near the center of the blade.

 40.1 **Reticulate** – Quinternaries anastomose with other veins to form polygons.

 40.1.1 **Regular** – Angles formed by vein intersections are regular (Fig. 289).

 40.1.2 **Irregular** – Angles formed by vein intersections are highly variable (Fig. 290).

 40.2 **Freely ramifying** – Quinternaries branch freely and are the finest vein-order the leaf exhibits (Fig. 291).

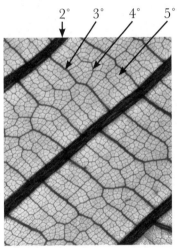

Fig. 289
Regular reticulate quinternaries
Pseudolmedia laevis
(Moraceae)

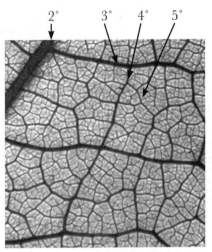

Fig. 290
Irregular reticulate quinternaries
Diospyros hispida
(Ebenaceae)

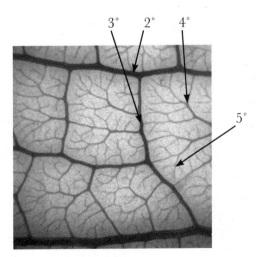

Fig. 291
Freely ramifying quinternaries
Stemonoporus nitidus
(Dipterocarpaceae)

41.   **Areolation** – Areoles are the smallest areas of the leaf tissue that are completely surrounded by veins; taken together they form a contiguous field of polygons over most of the area of the lamina. Any order of venation can form one or more sides of an areole.

    41.1   Lacking – Venation ramifies into the intercostal area without producing closed meshes (Fig. 292).

    41.2   Present

        41.2.1   Poor development – Areoles many-sided (often >7) and of highly irregular size and shape (Fig. 293).

        41.2.2   Moderate development – Areoles of irregular shape, more or less variable in size, generally with fewer sides than in poorly developed areolation (Fig. 294).

        41.2.3   Good development – Areoles of relatively consistent size and shape and generally with 3–6 sides (Fig. 295).

        41.2.4   Paxillate – Areoles occurring in distinct oriented fields (Fig. 296; definition is more general than in Hickey, 1979.)

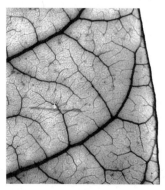 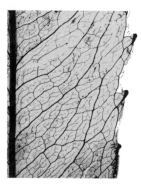 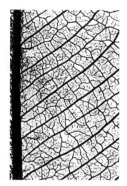

Fig. 292
Areolation lacking
Rhus taitensis
(Anacardiaceae)

Fig. 293
Areole development poor
Chloranthus glaber
(Chloranthaceae)

Fig. 294
Areole development moderate
Clusiella pendula
(Clusiaceae)

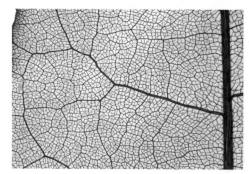 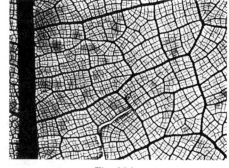

Fig. 295
Areole development good
Piranhea trifoliata
(Picrodendraceae)

Fig. 296
Areole development paxillate
Afrostyrax kamerunensis
(Huaceae)

42. **Freely Ending Veinlets (FEVs)** – Highest-order veins that freely ramify.

    42.1   FEV branching

        42.1.1 FEVs absent (Fig. 297).

        42.1.2 Mostly unbranched – FEVs present but unbranched, may be straight or curved (Fig. 298).

        42.1.3 Mostly with one branch (Fig. 299).

        42.1.4 Mostly with two or more branches

            42.1.4.1 Branching equal (dichotomous) (Fig. 300).

            42.1.4.2 Branching unequal (dendritic) (Fig. 301).

    42.2   FEV terminals

        42.2.1 Simple (Fig. 302).

        42.2.2 Tracheoid idioblasts – FEV endings are club-shaped and consist of tracheal cells with spiral wall thickenings (Foster, 1956; called dilated tracheal cells in Tucker, 1964) (Fig. 303).

        42.2.3 Highly branched sclereids – FEVs branch densely (10+) out of the plane of the veins; the finer branches often stain differently because they are sclereids, not tracheids (Fig. 304).

Fig. 297	Fig. 298	Fig. 299	Fig. 300	Fig. 301
FEVs absent	FEVs unbranched	FEVs one branched	FEVs dichotomous branching	FEVs dendritic branching

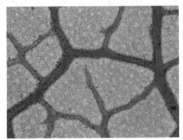 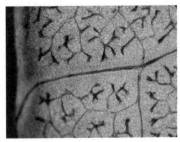 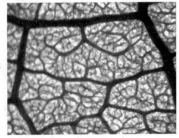

Fig. 302
Simple FEV terminals
Melicytus fasciger
(Violaceae)

Fig. 303
Tracheoid idioblasts
Bursera inaguensis
(Burseraceae)

Fig. 304
Highly branched sclereids
Tetragastris panamensis
(Burseraceae)

43.   **Marginal Ultimate Venation** – Configuration of the highest-order veins at the margin (see also II.29 on perimarginal veins)

43.1   **Absent** – Ultimate veins join perimarginal veins (Fig. 305).

43.2   **Incomplete** – Marginal ultimate veins recurve to form incomplete loops (Fig. 306).

43.3   **Spiked** – Marginal ultimate veins form outward-pointing spikes (Fig. 307).

43.4   **Looped** – Marginal ultimate vein recurved to form loops (Figs. 308, 309).

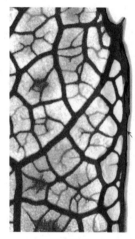

Fig. 305
Marginal ultimate venation absent
Pycnocoma littoralis
(Euphorbiaceae)

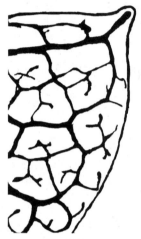

Fig. 306
Marginal ultimate veins
incomplete
(line drawing)

Fig. 307
Marginal ultimate veins
form spikes
(line drawing)

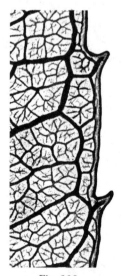

Fig. 308
Marginal ultimate venation looped
Mollinedia floribunda
(Monimiaceae)

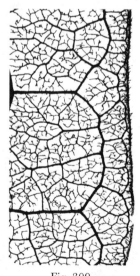

Fig. 309
Marginal ultimate venation looped
Picramnia krukovii
(Picramniaceae)

General Tooth Definitions

Leaf teeth contain a great number of systematically informative characters (Hickey and Wolfe, 1975; Hickey and Taylor, 1991; Doyle, 2007) and are extremely useful for circumscribing fossil leaf taxa. Their prevalence in fossil floras provides reliable proxy data about pre-Quaternary terrestrial paleotemperatures (Wolfe, 1971, 1995; Wilf, 1997; Utescher et al., 2000). Tooth size and shape appear to be useful variables for increasing precision in paleoclimate estimates and for paleoecological interpretation of fossil floras (Royer et al., 2005; Royer and Wilf, 2006).

Generally, a tooth can be recognized by its projection from the leaf margin (see I.13 and I.14) and its associated vasculature. Recognizing the boundaries of a tooth along the leaf margin can be difficult when sinuses are absent or teeth are widely separated. Some lab-tested, reproducible rules for defining tooth boundaries when high precision is necessary are found in Royer et al. (2005). Hickey and Taylor (1991) used tissue-level features to define *admedial* and *conjunctal* veins.

Definitions

distal flank
The portion of the margin between the tooth's apex and the nadir of the superjacent sinus (Fig. 310).

proximal flank
The portion of the margin between the tooth's apex and the sinus on the proximal side. The proximal sinus is recognized as the point where the curve of the tooth departs from the curve of the leaf margin, and may or may not coincide with the nadir of the subjacent sinus (Fig. 310).

sinus
A marginal embayment, incision, or indentation between marginal projections of any sort, typically lobes (Fig. 11), teeth (Figs. 11, 310) or the base of cordate leaves (Fig. 12)

tooth apex
The point of sharpest change in direction along the tooth margin, commonly but not always occurring at the most distal or exmedial point on the tooth (Fig. 310).

principal vein
The vein of widest gauge that enters the tooth (Fig. 311).

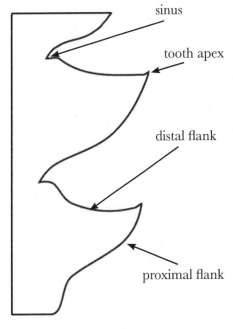

Fig. 310
The parts of a tooth

admedial vein

The first branch from the principal vein below the tooth apex that is of the same order or one order finer than the principal, and has >60% of its vascular tissue at its junction with the principal directed admedially or toward the mid-line of the leaf (Fig. 311).

accessory veins

All the veins between the tooth apex and the admedial vein that either branch from or merge with the principal vein. Typically the accessory veins of larger gauge have consistent courses in relation to the principal vein, admedial vein, and other tooth features, and such accessory veins commonly are conjunctal veins as defined below (Fig. 311).

conjunctal veins

Accessory veins that converge on or merge with the principal vein, contribute vascular tissue to the tooth apex, and have > 60% of their vascular tissue directed toward the tooth apex at their point of convergence or fusion with the principal vein. They may occur singly or in pairs that arise opposite or alternate to one another (Fig. 311).

gland

A discrete area of specialized cells that secrete by-products of plant metabolism. In fossils and cleared or dried leaves, the glands typically appear darker than the surrounding tissue. In addition to occurring on the lamina and petiole, they may occur in or be attached to the apex of the tooth (Fig. 312).

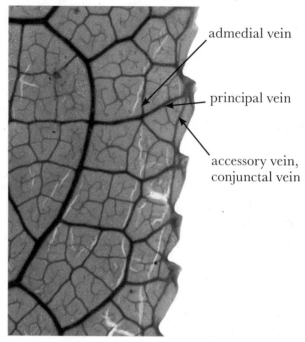

admedial vein

principal vein

accessory vein, conjunctal vein

Fig. 311
Aporusa frutescens
(Phyllanthaceae)

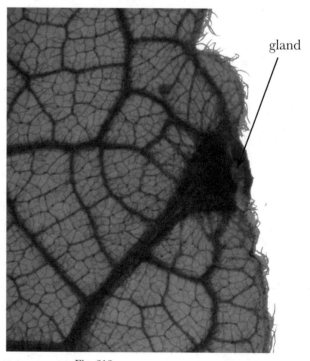

gland

Fig. 312
Gland
Gouania velutina
(Rhamnaceae)

III. Tooth Characters

44. **Tooth Spacing** – Distance between the corresponding points on adjacent teeth

44.1 **Regular** – Minimum intertooth distance is >60% of the maximum intertooth distance (Fig. 313).

44.2 **Irregular** – Minimum intertooth distance is <60% of the maximum intertooth distance (Fig. 314).

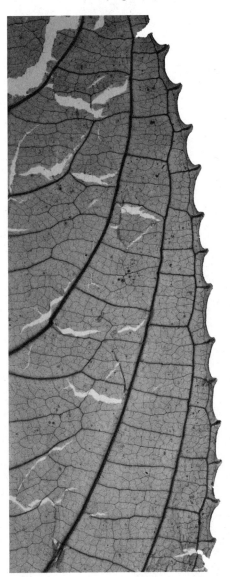

Fig. 313
Regular tooth spacing
Dichroa philippinensis
(Hydrangeaceae)

Fig. 314
Irregular tooth spacing
Campylostemon mucronatum
(Celastraceae)

45.   **Number of Orders of Teeth** – Number of discrete sizes of teeth. Sometimes, second- and third-order teeth occur in a regular series between first-order teeth.

    45.1   **One** – All teeth are the same size or vary in size continuously (Fig. 315).

    45.2   **Two** – Teeth are of two distinct sizes (Fig. 316).

    45.3   **Three** – Teeth are of three distinct sizes (Fig. 317).

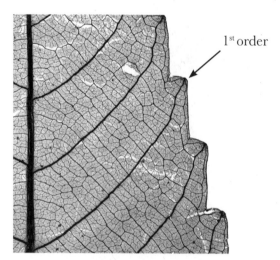

Fig. 315
One order of teeth
Leea macropus
(Vitaceae)

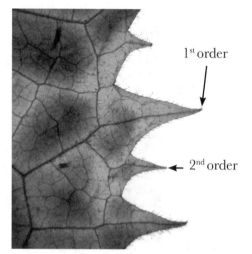

Fig. 316
Two orders of teeth
Aristotelia racemosa
(Elaeocarpaceae)

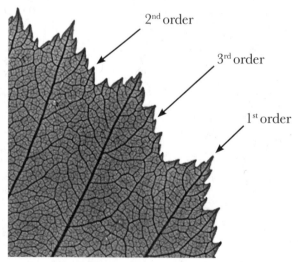

Fig. 317
Three orders of teeth
Crataegus brainerdii
(Rosaceae)

46.   **Number of Teeth per Centimeter** – Measured in the middle 50% of the leaf; that is, between 0.25 and 0.75 *L* (Fig. 318).

47.   **Sinus Shape**

47.1   **Angular** (Fig. 319).

47.2   **Rounded** (Fig. 320).

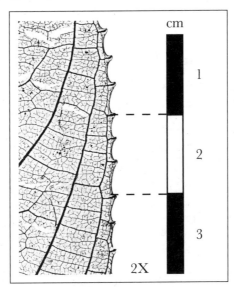

Fig. 318
Three teeth per cm
Dichroa philippinensis
(Hydrangeaceae)

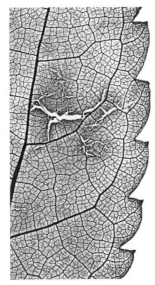

Fig. 319
Angular sinus
Celtis cerasifera
(Cannabaceae)

Fig. 320
Rounded sinus
Phylloclinium paradoxum
(Salicaceae)

48.   **Tooth Shape** – Described in terms of the distal and proximal flank curvatures relative to the midline of the tooth. The following states and abbreviations are used: convex (cv), straight (st), concave (cc), flexuous (fl; tooth flank is apically concave and basally convex), and retroflexed (rt; tooth flank is basally concave and apically convex). The distal flank shape is given first: for example, cc/fl indicates that the tooth is concave on the distal flank and flexuous on the proximal flank. The 25 possible combinations are shown in Figure 321 below. Note that a given leaf often exhibits more than one tooth shape.

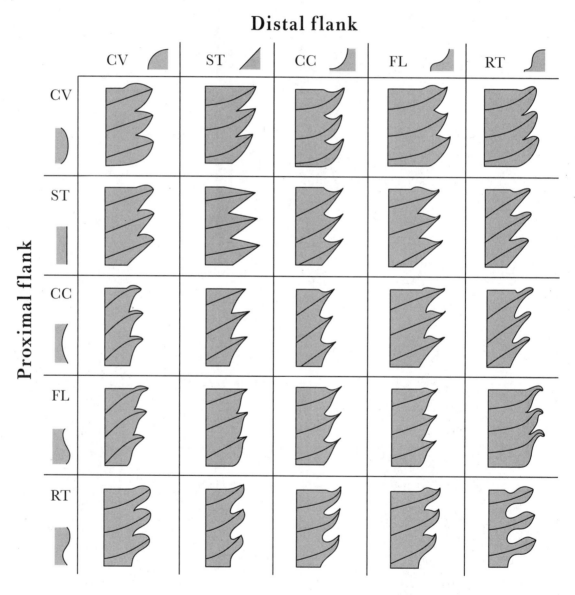

Fig. 321
Chart of possible tooth shapes. Always list the distal flank first.

49.   Principal Vein

49.1   Present (Figs. 322, 323, 324).

49.2   Absent – Generally occurs in teeth that are supplied by two or more veins of equal gauge (Fig. 325).

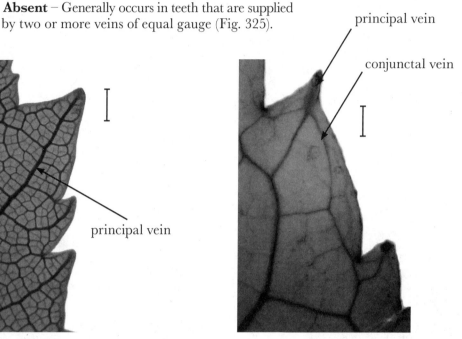

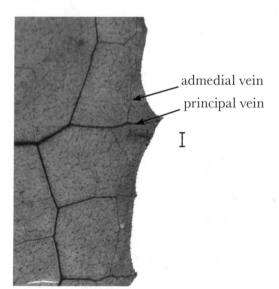

Fig. 322
Principal vein present
Carpinus laxiflora
(Betulaceae)
scale bar = 1 mm

Fig. 323
Principal vein present
Chloranthus serratus
(Chloranthaceae)
scale bar = 1 mm

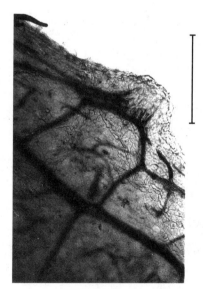

Fig. 324
Principal vein present
Martynia annua
(Martyniaceae)
scale bar = 1 mm

Fig. 325
Principal vein absent
Lopesia lopezoides
(Onagraceae)
scale bar = 100 μm

50. Principal Vein Termination

50.1 Submarginal (Fig. 326).

50.2 Marginal

50.2.1 At the apex of tooth (Fig. 327).

50.2.2 On the distal flank (Fig. 328).

50.2.3 At the nadir of superjacent sinus (Fig. 329).

50.2.4 On the proximal flank (Fig. 330).

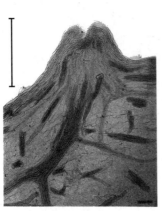

Fig. 326
Principal vein termination
submarginal
Fuchsia decidua
(Onagraceae)
scale bar = 100 μm

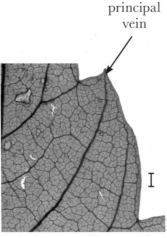

principal
vein

Fig. 327
Principal vein terminates
at the tooth apex
Acer negundo
(Sapindaceae)
scale bar = 1 mm

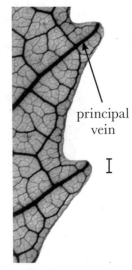

principal
vein

Fig. 328
Principal vein terminates on
the distal flank
Cupania vernalis
(Sapindaceae)
scale bar = 1 mm

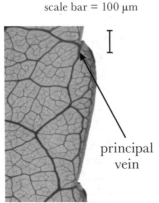

principal
vein

Fig. 329
Principal vein terminates at
nadir of the superjacent sinus
Elaeodendron glaucum
(Celastraceae)
scale bar = 1 mm

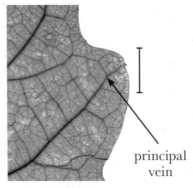

principal
vein

Fig. 330
Principal vein terminates
on the proximal flank
Quercus alba × *velutina*
(Fagaceae)
scale bar = 5 mm

51. Course of Major Accessory Vein(s)

51.1 Convex relative to principal vein (Fig. 331).

51.1.1 Looped – With multiple looping connections to principal vein (Fig. 332).

51.2 Straight or concave to principal vein (Figs. 333, 334).

51.3 Running from sinus to principal vein (Fig. 335).

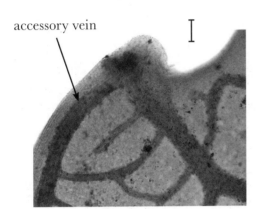

accessory vein

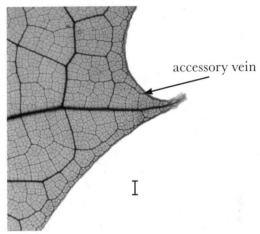

accessory vein

Fig. 331
Accessory veins convex
Melicytus fasciger
(Violaceae)
scale bar = 100 μm

Fig. 332
Accessory veins looped
Platanus orientalis
(Platanaceae)
scale bar = 1 mm

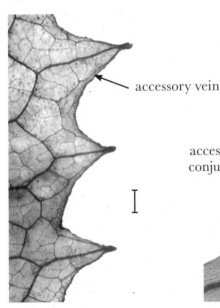

accessory vein

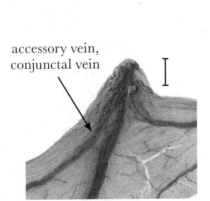

accessory vein,
conjunctal vein

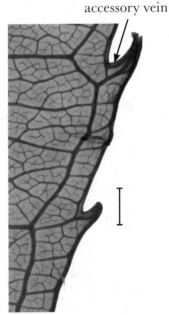

accessory vein

Fig. 333
Accessory vein straight
Diphylleia grayi
(Berberidaceae)
scale bar = 1 mm

Fig. 334
Accessory vein concave
Vitis cavaleriei
(Vitaceae)
scale bar = 100 μm

Fig. 335
Accessory vein running from sinus
Vitis cavaleriei
(Vitacea)
scale bar = 1 mm

52. Special Features of the Tooth Apex

52.1   Simple – No tissue or structure is present within or on the tooth apex (Fig. 336).

52.2   Specific tissue or structure present within the tooth apex

   52.2.1   Foraminate – Having a bulb- or funnel-shaped cavity at the tooth apex that opens to the outside (Fig. 337).

   52.2.2   Tylate – Having clear tissue at the termination of the principal vein (Fig. 338).

   52.2.3   Cassidate – Having opaque tissue at the termination of the principal vein (Fig. 339).

52.3   Specific tissue or structure on the tooth apex

   52.3.1   Spinose – Principal vein extends beyond the leaf margin; extension may be short or long, usually sharp (Fig. 340).

   52.3.2   Mucronate – An opaque, vascularized, peg-shaped, non-deciduous projection is present at the apex (Fig. 341).

   52.3.3   Setaceous – An opaque, peg-shaped, deciduous projection is present at the apex (Fig. 342).

   52.3.4   Papillate – A clear, flame-shaped projection is present at the apex (Fig. 343).

   52.3.5   Spherulate – A clear, spherical projection is present at the apex (Fig. 344).

52.4   Nonspecific – In fossils, it is often not possible to distinguish the type of gland or structures at the tooth apex. This character state can be used for the description of fossil teeth with a visible concentration of material in or on the tooth apex not assignable to the categories above (Fig. 345).

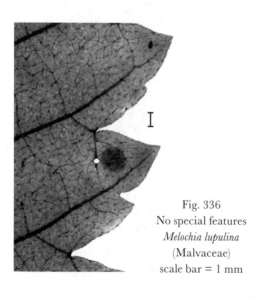

Fig. 336
No special features
Melochia lupulina
(Malvaceae)
scale bar = 1 mm

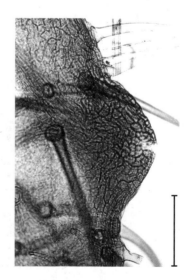

Fig. 337
Foraminate tooth apex
Circaea erubescens
(Onagraceae)
scale bar = 100 µm

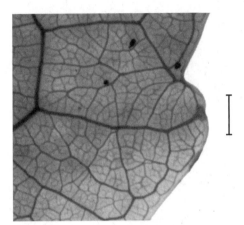

Fig. 338
Tylate tooth apex
Homalium racemosum
(Salicaceae)
scale bar = 1 mm

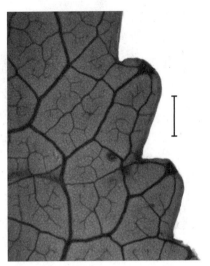

Fig. 339
Cassidate tooth apex
Tetracentron sinense
(Trochodendraceae)
scale bar = 1 mm

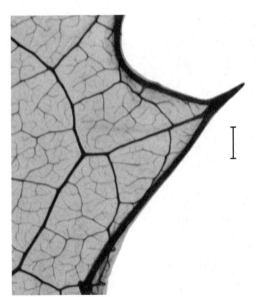

Fig. 340
Spinose tooth apex
Ilex dipyrena
(Aquifoliaceae)
scale bar = 1 mm

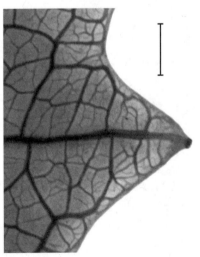

Fig. 341
Mucronate tooth apex
Trimeria alnifolia
(Salicaceae)
scale bar = 1 mm

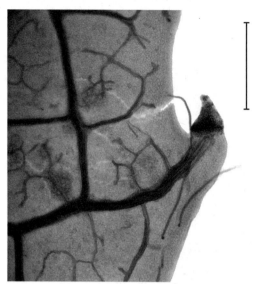

Fig. 342
Setaceous tooth apex
Thea sinensis
(Theaceae)
scale bar = 1 mm

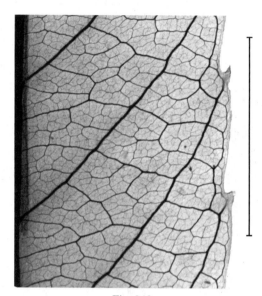

Fig. 343
Papillate tooth apex
Schumacheria castaneifolia
(Dilleniaceae)
scale bar = 10 mm

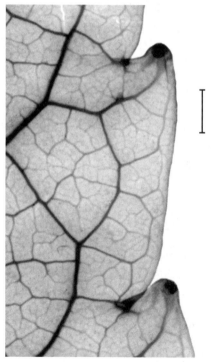

Fig. 344
Spherulate tooth apex
Idesia polycarpa
(Salicaceae)
scale bar = 1 mm

Fig. 345
Nonspecific tooth apex (fossil)
Cercidiphyllum genetrix
(Cercidiphyllaceae)
scale bar = 5 mm

Appendix A. Outline of Characters and Character States

I. Leaf Characters

1. Leaf Attachment
  1.1 Petiolate
  1.2 Sessile

2. Leaf Arrangement
  2.1 Alternate
  2.2 Subopposite
  2.3 Opposite
  2.4 Whorled

3. Leaf Organization
  3.1 Simple
  3.2 Compound
    3.2.1 Palmately compound
    3.2.2 Pinnately compound
      3.2.2.1 Once
      3.2.2.2 Twice
      3.2.2.3 Thrice

4. Leaflet Arrangement
  4.1 Alternate
  4.2 Subopposite
  4.3 Opposite
    4.3.1 Odd-pinnately compound
    4.3.2 Even-pinnately compound
  4.4 Unknown

5. Leaflet Attachment
  5.1 Petiolulate
  5.2 Sessile

6. Petiol(ul)e Features
  6.1 Petiol(ul)e base
    6.1.1 Sheathing
    6.1.2 Pulvin(ul)ate
  6.2 Glands
    6.2.1 Petiolar
    6.2.2 Acropetiolar
  6.3 Petiole cross-section
    6.3.1 Terete
    6.3.2 Semi-terete
    6.3.3 Canaliculate
    6.3.4 Triangular
    6.3.5 Alate
  6.4 Phyllodes

7. Position of Lamina Attachment
  7.1 Marginal
  7.2 Peltate central
  7.3 Peltate excentric

8. Laminar Size
  8.1 Leptophyll
  8.2 Nanophyll
  8.3 Microphyll
  8.4 Notophyll
  8.5 Mesophyll
  8.6 Macrophyll
  8.7 Megaphyll

9. Laminar L:W Ratio

10. Laminar Shape
  10.1 Elliptic
  10.2 Obovate
  10.3 Ovate
  10.4 Oblong
  10.5 Linear
  10.6 Special

11. Medial Symmetry
  11.1 Symmetrical
  11.2 Asymmetrical

12. Base Symmetry
  12.1 Symmetrical
  12.2 Asymmetrical
    12.2.1 Basal width asymmetrical
    12.2.2 Basal extension asymmetrical
    12.2.3 Basal insertion asymmetrical

13. Lobation
  13.1 Unlobed
  13.2 Lobed
    13.2.1 Palmately lobed
      13.2.1.1 Palmatisect
    13.2.2 Pinnately lobed
      13.2.2.1 Pinnatisect
    13.2.3 Palmately and pinnately lobed
    13.2.4 Bilobed

14. Margin Type
  14.1 Untoothed
  14.2 Toothed
    14.2.1 Dentate
    14.2.2 Serrate
    14.2.3 Crenate

15. Special Margin Features
  15.1 Appearance of the edge of the blade
    15.1.1 Erose
    15.1.2 Sinuous
  15.2 Appearance of the plane of the blade
    15.2.1 Revolute
    15.2.2 Involute
    15.2.3 Undulate

16. Apex Angle
  16.1 Acute
  16.2 Obtuse
  16.3 Reflex

17. Apex Shape
  17.1 Straight
  17.2 Convex
    17.2.1 Rounded
    17.2.2 Truncate
  17.3 Acuminate
  17.4 Emarginate
  17.5 Lobed

18. Base Angle
  18.1 Acute
  18.2 Obtuse
  18.3 Reflex
  18.4 Circular

19. Base Shape
  19.1 $l_b = 0$
    19.1.1 Straight (cuneate)
    19.1.2 Concave
    19.1.3 Convex
      19.1.3.1 Rounded
      19.1.3.2 Truncate
    19.1.4 Concavo-convex
    19.1.5 Complex
    19.1.6 Decurrent
  19.2 $l_b > 0$ or $l_b \sim 0$
    19.2.1 Cordate
    19.2.2 Lobate
      19.2.2.1 Sagittate
      19.2.2.2 Hastate
      19.2.2.3 Runcinate
      19.2.2.4 Auriculate

20. Terminal Apex Features
  20.1 Mucronate
  20.2 Spinose
  20.3 Retuse

21. Surface Texture
  21.1 Smooth
  21.2 Pitted
  21.3 Papillate
  21.4 Rugose
  21.5 Pubescent

22. Surficial Glands
  22.1 Laminar
  22.2 Marginal
  22.3 Apical
  22.4 Basal laminar

II. Vein Characters

23. Primary Vein Framework
  23.1 Pinnate
  23.2 Palmate
    23.2.1 Actinodromous
      23.2.1.1 Basal
      23.2.1.2 Suprabasal
    23.2.2 Palinactinodromous
    23.2.3 Acrodromous
      23.2.3.1 Basal
      23.2.3.2 Suprabasal
    23.2.4 Flabellate
    23.2.5 Parallelodromous
    23.2.6 Campylodromous

24. Naked Basal Veins
  24.1 Absent
  24.2 Present

25. Number of Basal Veins

26. Agrophic Veins
  26.1 Absent
  26.2 Present
    26.2.1 Simple
    26.2.2 Compound

27. Major 2° Vein Framework
  27.1 Major secondaries reach margin
    27.1.1 Craspedodromous
    27.1.2 Semicraspedodromous
    27.1.3 Festooned semicraspedodromous

27.2 Major secondaries do not reach margin
    and lose gauge by attenuation
  27.2.1 Eucamptodromous
    27.2.1.1 Basal eucamptodromous
    27.2.1.2 Hemieucamtodromous
    27.2.1.3 Eucamptodromous
        becoming brochidodromous
        distally
  27.2.2 Reticulodromous
  27.2.3 Cladodromous
27.3 Major secondaries form loops of 2°
    gauge and do not reach margin.
  27.3.1 Simple brochidodromous
  27.3.2 Festooned brochidodromous
27.4 Mixed

28. Interior Secondaries
  28.1 Absent
  28.2 Present

29. Minor Secondary Course
  29.1 Craspedodromous
  29.2 Simple brochidodromous
  29.3 Semicraspedodromous

30. Perimarginal Veins
  30.1 Marginal secondary
  30.2 Intramarginal secondary
  30.3 Fimbrial vein

31. Major Secondary Spacing
  31.1 Regular
  31.2 Irregular
  31.3 Decreasing proximally
  31.4 Gradually increasing proximally
  31.5 Abruptly increasing proximally

32. Variation of Major Secondary
    Angle to Midvein
  32.1 Uniform
  32.2 Inconsistent
  32.3 Smoothly increasing proximally
  32.4 Smoothly decreasing proximally
  32.5 Abruptly increasing proximally
  32.6 One pair acute basal secondaries

33. Major Secondary Attachment to Midvein
  33.1 Decurrent
  33.2 Proximal secondaries decurrent
  33.3 Excurrent
  33.4 Deflected

34. Intersecondary Veins
  34.1 Intersecondary proximal course
    34.1.1 Parallel to major secondaries
    34.1.2 Perpendicular to midvein
  34.2 Intersecondary length
    34.2.1 Less than 50% of subjacent
        secondary
    34.2.2 More than 50% of subjacent
        secondary
  34.3 Intersecondary distal course
    34.3.1 Reticulating or ramifying
    34.3.2 Parallel to major secondary
    34.3.3 Perpendicular to subjacent major
        secondary
    34.3.4 Basiflexed, not joining subjacent
        secondary at right angle
  34.4 Intersecondary frequency
    34.4.1 Less than 1 per intercostal area
    34.4.2 Usually 1 per intercostal area
    34.4.3 More than 1 per intercostal area

35. Intercostal Tertiary Vein Fabric
  35.1 Percurrent
    35.1.1 Course of percurrent tertiaries
      35.1.1.1 Opposite
        35.1.1.1.1 Straight
        35.1.1.1.2 Convex
        35.1.1.1.3 Sinuous
        35.1.1.1.4 Chevroned
      35.1.1.2 Alternate
      35.1.1.3 Mixed opposite-alternate
    35.1.2 Angle of percurrent tertiaries
      35.1.2.1 Acute
      35.1.2.2 Obtuse
      35.1.2.3 Perpendicular
  35.2 Reticulate
    35.2.1 Irregular
    35.2.2 Regular
    35.2.3 Composite admedial
  35.3 Ramified
    35.3.1 Admedially ramified
    35.3.2 Exmedially ramified
    35.3.3 Transverse ramified
    35.3.4 Transverse freely ramified

36. Intercostal Tertiary Vein Angle Variability
    36.1 Inconsistent
    36.2 Consistent
    36.3 Increasing exmedially
        36.3.1 Basally concentric
    36.4 Decreasing exmedially
    36.5 Increasing proximally
    36.6 Decreasing proximally

37. Epimedial Tertiaries
    37.1 Epimedial tertiary fabric
        37.1.1 Percurrent
            37.1.1.1 Opposite
            37.1.1.2 Alternate
            37.1.1.3 Mixed
        37.1.2 Ramified
        37.1.3 Reticulate
        37.1.4 Mixed
    37.2 Course of percurrent epimedial
        tertiaries
        37.2.1 Admedial course
            37.2.1.1 Parallel to subjacent
                secondary
            37.2.1.2 Parallel to intercostal
                tertiaries
            37.2.1.3 Perpendicular to midvein
            37.2.1.4 Parallel to intersecondary
            37.2.1.5 Obtuse to midvein
            37.2.1.6 Acute to midvein
        37.2.2 Exmedial course
            37.2.2.1 Parallel to intercostal
                tertiary
            37.2.2.2 Basiflexed
            37.2.2.3 Acroflexed

38. Exterior Tertiary Course
    38.1 Absent
    38.2 Looped
    38.3 Terminating at margin
    38.4 Variable

39. Quaternary Vein Fabric
    39.1 Percurrent
        39.1.1 Opposite
        39.1.2 Alternate
        39.1.3 Mixed percurrent
    39.2 Reticulate
        39.2.1 Regular
        39.2.2 Irregular
    39.3 Freely ramifying

40. Quinternary Vein Fabric
    40.1 Reticulate
        40.1.1 Regular
        40.1.2 Irregular
    40.2 Freely ramifying

41. Areolation
    41.1 Lacking
    41.2 Present
        41.2.1 Poor development
        41.2.2 Moderate development
        41.2.3 Good development
        41.2.4 Paxillate

42. Freely Ending Veinlets (FEVs)
    42.1 FEV branching
        42.1.1 FEVs absent
        42.1.2 Mostly unbranched
        42.1.3 Mostly 1-branched
        42.1.4 Mostly 2- or more branched
            42.1.4.1 Branching equal
                (dichotomous)
            42.1.4.2 Branching unequal
                (dendritic)
    42.2 FEV terminals
        42.2.1 Simple
        42.2.2 Tracheoid idioblasts
        42.2.3 Highly branched sclereids

43. Marginal Ultimate Venation
    43.1 Absent
    43.2 Incomplete loops
    43.3 Spiked
    43.4 Looped

III. Tooth Characters

44. Tooth Spacing
    44.1 Regular
    44.2 Irregular

45. Number of Orders of Teeth
    45.1 One
    45.2 Two
    45.3 Three

46. Number of Teeth/cm

47. Sinus Shape
    47.1 Angular
    47.2 Rounded

48. Tooth Shape (cv, st, cc, fl, rt)
    (distal flank listed first)

49. Principal Vein
    49.1 Present
    49.2 Absent

50. Principal Vein Termination
    50.1 Submarginal
    50.2 Marginal
        50.2.1 At apex of tooth
        50.2.2 On distal flank
        50.2.3 At nadir of superjacent sinus
        50.2.4 On proximal flank

51. Course of Ancillary Veins Relative to
        Principal Vein
    51.1 Convex
        51.1.1 Looped
    51.2 Straight or concave
    51.3 Running from sinus

52. Special Features of the Tooth Apex
    52.1 None
    52.2 Within tooth apex
        52.2.1 Foraminate
        52.2.2 Tylate
        52.2.3 Cassidate
    52.3 On tooth apex
        52.3.1 Spinose
        52.3.2 Mucronate
        52.3.3 Setaceous
        52.3.4 Papillate
        52.3.5 Spherulate
    52.4 Nonspecific

Appendix B. Examples of Fully Described Leaves with Images

The eighteen examples in this appendix are keyed to the numeric codes for each character state described in the text. The easiest way to score leaves or review the scores that we assigned to these examples is to photocopy Appendix A and use it as a guide to all of the possible character states. These numeric codes can also be used to quickly and fully describe a leaf's characteristics in a computer database.

The examples are scored in a Microsoft® Excel® 2007 worksheet, shown on the facing page. Worksheets can be downloaded from http://www.paleobotanyproject.org/. When the number of the appropriate character state is typed into the "score" column, the description field is populated automatically. (Note: The shaded boxes are skipped and the actual values for these character states are typed into the description field.) Once the worksheet is completed, the user can insert a verbal description of the leaf in the lower right-hand corner or upload the data into a database. The blank template on the facing page can be used to quickly capture the numeric codes. Two boxes are provided for a character state when a range of choices is needed. Because the extant leaf images in Appendix B do not illustrate the leaf attachment and organization characters, the first six characters were scored by reviewing herbarium sheets.

Worksheets can be downloaded from http://www.paleobotanyproject.org/.

To improve data entry, we use the following generic codes:

0 = Absent – The character is not present in this leaf. For example, *Tilia mandshurica* (Appendix B, example 1) does not have intersecondary veins, so this character is absent.

88 = Not visible – This character is not preserved and so cannot be scored.

99 = Not applicable (n/a) – The character does not apply to this leaf. For example, tooth type would score as n/a for a leaf that has a smooth margin.

Excel Leaf Scoring Template

I. Leaf Characters	Score	Description
Leaf Attachment		
Leaf Arrangement		
Leaf Organization		
Leaflet Arrangement		
Leaflet Attachment		
Petiole Features		

Features of the Blade

Position of Blade Attachment		
Laminar Size		
Laminar L:W Ratio		
Laminar Shape		
Medial Symmetry		
Base Symmetry		
Base Symmetry		
Lobation		
Margin Type		
Special Margin Features		
Apex Angle		
Apex Shape		
Base Angle		
Base Shape		
Base Shape		
Terminal Apex Features		
Surface Texture		
Surficial Glands		

II. Venation		Score	Description
1°	Primary Vein Framework		
	Naked Basal Veins		
	Number of Basal Veins		
	Agrophic Veins		
2°	Major 2° Vein Framework		
	Interior Secondaries		
	Minor Secondary Course		
	Perimarginal Veins		
	Major Secondary Spacing		
	Variation of Secondary Angle		
	Major Secondary Attachment		
Inter-2°	Proximal Course		
	Length		
	Distal Course		
	Vein Frequency		
3°	Intercostal 3° Vein Fabric		
	Angle of Percurrent Tertiaries		
	Vein Angle Variability		
	Epimedial Tertiaries		
	Admedial Course		
	Exmedial Course		
	Exterior Tertiary Course		
4°	Quaternary Vein Fabric		
5°	Quinternary Vein Fabric		
	Areolation		
	FEV branching		
	FEV termination		
	Marginal Ultimate Venation		

III. Teeth	Score	Description
Tooth Spacing		
Number of Orders of Teeth		
Teeth / cm		
Sinus Shape		
Tooth Shapes		
Tooth Shapes		
Tooth Shapes		
Tooth Shapes		
Principal Vein		
Principal Vein Termination		
Course of Accessory Vein		
Features of the Tooth Apex		

Text Description:

Example 1. Malvaceae - *Tilia baccata* var. *mandshurica*

Malvaceae - *Tilia mandshurica*

I. Leaf Characters	Score	Description
Leaf Attachment	1.1	petiolate
Leaf Arrangement	2.1	alternate
Leaf Organization	3.1	simple
Leaflet Arrangement	99	n/a
Leaflet Attachment	99	n/a
Petiole Features	88	not visible

Features of the Blade

	Score	Description
Position of Blade Attachment	7.1	marginal
Laminar Size	8.5	mesophyll
Laminar L:W Ratio		1.2:1
Laminar Shape	10.3	ovate
Medial Symmetry	11.1	symmetrical
Base Symmetry	12.2.1	extension asymmetry
Base Symmetry	12.2.1	extension asymmetry
Lobation	13.1	unlobed
Margin Type	14.2.2	serrate
Special Margin Features	0	absent
Apex Angle	16.1	acute
Apex Shape	17.3	acuminate
Base Angle	18.2	obtuse
Base Shape	19.2.1	cordate
Base Shape	19.2.1	cordate
Terminal Apex Features	88	not visible
Surface Texture	88	not visible
Surficial Glands	0	absent

	II. Venation	Score	Description
1°	Primary Vein Framework	23.2.1 .1	basal actinodromous
	Naked Basal Veins	0	absent
	Number of Basal Veins		8
	Agrophic Veins	26.2	compound
2°	Major 2° Vein Framework	27.1.2	semicraspedodromous
	Interior Secondaries	0	absent
	Minor Secondary Course	29.1	craspedodromous
	Perimarginal Veins	30	absent
	Major Secondary Spacing	31.4	gradually increasing proximally
	Variation of Secondary Angle	32.1	uniform
	Major Secondary Attachment	33.3	excurrent
Inter-2°	Proximal Course	0	absent
	Length	0	absent
	Distal Course	0	absent
	Vein Frequency	0	absent
3°	Intercostal 3° Vein Fabric	35.1.1.1.2	convex opposite percurrent
	Angle of Percurrent Tertiaries	35.1.2.2	obtuse to midvein
	Vein Angle Variability	36.3.1	basally concentric
	Epimedial Tertiaries	37.1.1.1	opposite percurrent
	Admedial Course	37.2.1.6	acute
	Exmedial Course	37.2.2.1	parallel to intercostal tertiary
	Exterior Tertiary Course	38.4	variable
4°	Quaternary Vein Fabric	39.1.2	alternate percurrent
5°	Quinternary Vein Fabric	40.1.1	regular reticulate
	Areolation	41.2.3	good development
	FEV branching	88	not visible
	FEV termination	88	not visible
	Marginal Ultimate Venation	88	not visible

III. Teeth	Score	Description
Tooth Spacing	44.1	regular
Number of Orders of Teeth	45.1	one
Teeth / cm		2
Sinus Shape	47.2	rounded
Tooth Shapes		cc/st
Tooth Shapes		cc/cc
Tooth Shapes		cc/fl
Tooth Shapes		
Principal Vein	49.1	present
Principal Vein Termination	50.2.1	at apex of tooth
Course of Accessory Vein	99	n/a
Features of the Tooth Apex	52.1	none

Text Description:

Leaf attachment petiolate. Marginal blade attachment. Laminar size mesophyll with L:W ratio of 1.2:1. Laminar shape elliptic to ovate, symmetrical with basal extension asymmetry. Margin unlobed with serrate teeth. Apex angle acute with acuminate shape. Base angle obtuse with cordate shape. Primary veins basal actinodromous with eight basal veins. Compound agrophic veins present. Major secondary framework semicraspedodromous, minor secondaries craspedodromous, major secondary spacing gradually increasing proximally with uniform angle and excurrent attachment. Intersecondaries absent. Intercostal tertiary fabric opposite percurrent with convex course, obtuse angle to midvein, with basally concentric tertiaries. Epimedial tertiaries opposite percurrent with acute admedial course, and exmedial course parallel to intercostal tertiary. Exterior tertiary course variable. Quaternary vein fabric alternate percurrent. Quinternary vein fabric regular reticulate. Areolation shows good development but FEVs are not visible. Tooth spacing regular, with a single order of teeth. Sinus shape rounded with tooth shapes: concave/straight, concave/concave, and concave/flexuous. Principal vein terminates at apex of tooth. Accessory veins absent.

Example 2. Dilleniaceae - *Davilla rugosa*

Dilleniaceae - *Davilla rugosa*

I. Leaf Characters	Score	Description
Leaf Attachment	1.1	petiolate
Leaf Arrangement	2.1	alternate
Leaf Organization	3.1	simple
Leaflet Arrangement	99	n/a
Leaflet Attachment	99	n/a
Petiole Features	88	not visible

Features of the Blade

	Score	Description
Position of Blade Attachment	7.1	marginal
Laminar Size	8.4	notophyll
Laminar L:W Ratio		1.8:1
Laminar Shape	10.1	elliptic
Medial Symmetry	11.1	symmetrical
Base Symmetry	12.1	symmetrical
Base Symmetry	12.1	symmetrical
Lobation	13.1	unlobed
Margin Type	14.1	untoothed
Special Margin Features		not visible
Apex Angle	16.2	obtuse
Apex Shape	17.2.1	rounded
Base Angle	18.2	obtuse
Base Shape	19.1.3.1	rounded
Base Shape	19.1.3.1	rounded
Terminal Apex Features	0	absent
Surface Texture	88	not visible
Surficial Glands	88	not visible

II. Venation		Score	Description
1°	Primary Vein Framework	23.1	pinnate
	Naked Basal Veins	24.1	absent
	Number of Basal Veins		1
	Agrophic Veins	0	absent
2°	Major 2° Vein Framework	27.3.1	simple brochidodromous
	Interior Secondaries	28.1	absent
	Minor Secondary Course	0	absent
	Perimarginal Veins	30.3	fimbrial vein
	Major Secondary Spacing	31.1	regular
	Variation of Secondary Angle	32.1	uniform
	Major Secondary Attachment	33.3	excurrent
Inter-2°	Proximal Course	0	absent
	Length	99	n/a
	Distal Course	99	n/a
	Vein Frequency	99	n/a
3°	Intercostal 3° Vein Fabric	35.1.1.3	sinuous opposite percurrent
	Angle of Percurrent Tertiaries	35.1.2.2	obtuse to midvein
	Vein Angle Variability	36.5	increasing exmedially
	Epimedial Tertiaries	37.1.1.1	opposite percurrent
	Admedial Course	37.2.1.3	perpendicular to midvein
	Exmedial Course	37.2.2.1	parallel to intercostal tertiary
	Exterior Tertiary Course	38.2	looped
4°	Quaternary Vein Fabric	39.2.2	irregular reticulate
5°	Quinternary Vein Fabric	40.1.2	irregular reticulate
	Areolation	41.2.2	moderate development
	FEV branching	42.1.4.2	2 or more, dendritic
	FEV termination	42.2.1	simple
	Marginal Ultimate Venation	43.3	looped

III. Teeth	Score	Description
Tooth Spacing	99	n/a
Number of Orders of Teeth	99	n/a
Teeth / cm	99	n/a
Sinus Shape	99	n/a
Tooth Shapes	99	n/a
Tooth Shapes	99	n/a
Tooth Shapes	99	n/a
Tooth Shapes	99	n/a
Principal Vein	99	n/a
Principal Vein Termination	99	n/a
Course of Accessory Vein	99	n/a
Features of the Tooth Apex	99	n/a

Text Description:

Blade attachment marginal. Laminar size notophyll, L:W ratio 1.8:1, laminar shape elliptic with medial symmetry and basal symmetry. Margin is entire with obtuse apex angle, rounded apex, obtuse base angle, and rounded base shape. Primary venation pinnate with no naked basal veins, one basal vein, and no agrophic veins. Major secondaries simple brochidodromous with regular spacing, uniform angle and excurrent attachment to midvein. Interior secondaries absent, minor secondaries absent, intersecondaries absent, fimbrial vein present. Intercostal tertiary veins mixed percurrent with obtuse angle that increases exmedially. Epimedial tertiaries opposite percurrent with proximal course perpendicular to the midvein and distal course parallel to the intercostal tertiaries. Exterior tertiaries looped. Quaternary vein fabric irregular reticulate. Quinternary vein fabric irregular reticulate. Areolation shows moderate development. Freely ending veinlets have two or more dendritic branches, and marginal ultimate venation is looped.

Example 3. Dipterocarpaceae - *Stemonoporus nitidus*

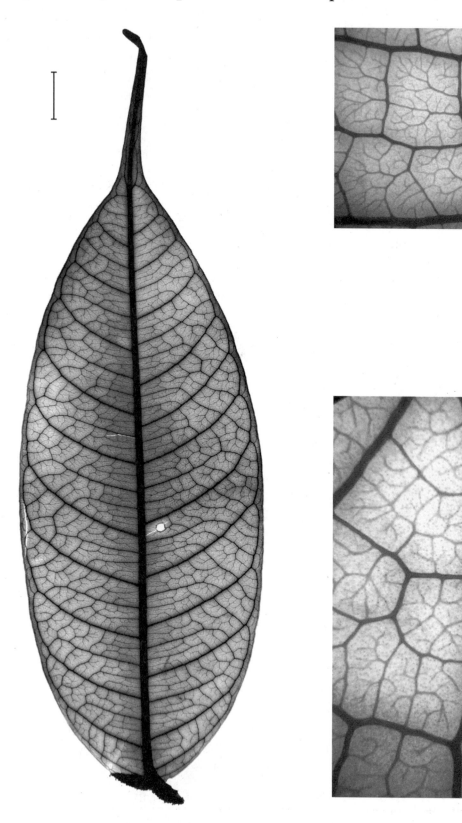

Dipterocarpaceae - *Stemonoporus nitidus*

I. Leaf Characters	Score	Description
Leaf Attachment	1.1	petiolate
Leaf Arrangement	2.1	alternate
Leaf Organization	3.1	simple
Leaflet Arrangement	99	n/a
Leaflet Attachment	99	n/a
Petiole Features	88	not visible

Features of the Blade

	Score	Description
Position of Blade Attachment	7.1	marginal
Laminar Size	8.4	notophyll
Laminar L:W Ratio		3:1
Laminar Shape	10.1	elliptic
Medial Symmetry	11.1	symmetrical
Base Symmetry	12.1	symmetrical
Base Symmetry	12.1	symmetrical
Lobation	13.1	unlobed
Margin Type	14.1	untoothed
Special Margin Features		not visible
Apex Angle	16.1	acute
Apex Shape	17.3	acuminate
Base Angle	18.2	obtuse
Base Shape	19.1.3.1	rounded
Base Shape	19.1.3.1	rounded
Terminal Apex Features	0	absent
Surface Texture	88	not visible
Surficial Glands	88	not visible

	II. Venation	Score	Description
1°	Primary Vein Framework	23..1	pinnate
	Naked Basal Veins	24.1	absent
	Number of Basal Veins		1
	Agrophic Veins	26.1	absent
2°	Major 2° Vein Framework	27.3.1	simple brochidodromous
	Interior Secondaries	28.1	absent
	Minor Secondary Course	0	absent
	Perimarginal Veins	30.3	fimbrial vein
	Major Secondary Spacing	31.1	regular
	Variation of Secondary Angle	32.1	uniform
	Major Secondary Attachment	33.3	excurrent
Inter- 2°	Proximal Course	34.1.1	parallel to major secondaries
	Length	34.2.2	>50%
	Distal Course	34.3.4	basiflexed
	Vein Frequency	34.4.2	~1 per intercostal area
3°	Intercostal 3° Vein Fabric	35.1.1.3	mixed percurrent
	Angle of Percurrent Tertiaries	35.1.2.2	obtuse
	Vein Angle Variability	36.2	consistent
	Epimedial Tertiaries	37.1.1.1	opposite percurrent
	Admedial Course	37.2.1.3	perpendicular to midvein
	Exmedial Course	37.2.2.2	basiflexed
	Exterior Tertiary Course	38.2	looped
4°	Quaternary Vein Fabric	39.2.2	irregular reticulate
5°	Quinternary Vein Fabric	40.2	freely ramifying
	Areolation	41.2.2	moderate development
	FEV branching	42.1.4.2	2 or more, dendritic
	FEV termination	42.2.1	simple
	Marginal Ultimate Venation	43.1	absent

III. Teeth	Score	Description
Tooth Spacing	99	n/a
Number of Orders of Teeth	99	n/a
Teeth / cm	99	n/a
Sinus Shape	99	n/a
Tooth Shapes	99	n/a
Tooth Shapes	99	n/a
Tooth Shapes	99	n/a
Tooth Shapes	99	n/a
Principal Vein	99	n/a
Principal Vein Termination	99	n/a
Course of Accessory Vein	99	n/a
Features of the Tooth Apex	99	n/a

Text Description:

Leaf attachment petiolate. Blade attachment marginal, laminar size notophyll, L:W ratio 3:1, laminar shape elliptic with medial symmetry and basal symmetry. Margin entire with acute apex angle, acuminate apex, obtuse base angle, and rounded base shape. Primary venation pinnate with no naked basal veins, one basal vein, and no agrophic veins. Major secondaries simple brochidodromous with regular spacing, uniform angle, and excurrent attachment to midvein. Interior secondaries absent, minor secondaries absent, and fimbrial vein present. Intersecondaries span more than 50% of the length of the subjacent secondary, occur at slightly more than one per intercostal area, proximal course is parallel to major secondaries, and distal course is basiflexed and parallel to intercostal tertiaries. Intercostal tertiary veins mixed percurrent with obtuse angle that remains consistently. Epimedial tertiaries opposite percurrent with proximal course perpendicular to the midvein and distal course basiflexed. Exterior tertiaries looped. Quaternary vein fabric irregular reticulate. Quinternary vein fabric freely ramifying. Areolation shows moderate development. Freely ending veinlets mostly two branched and marginal ultimate venation joins fimbrial vein.

Example 4. Fabaceae - *Bauhinia madagascariensis*

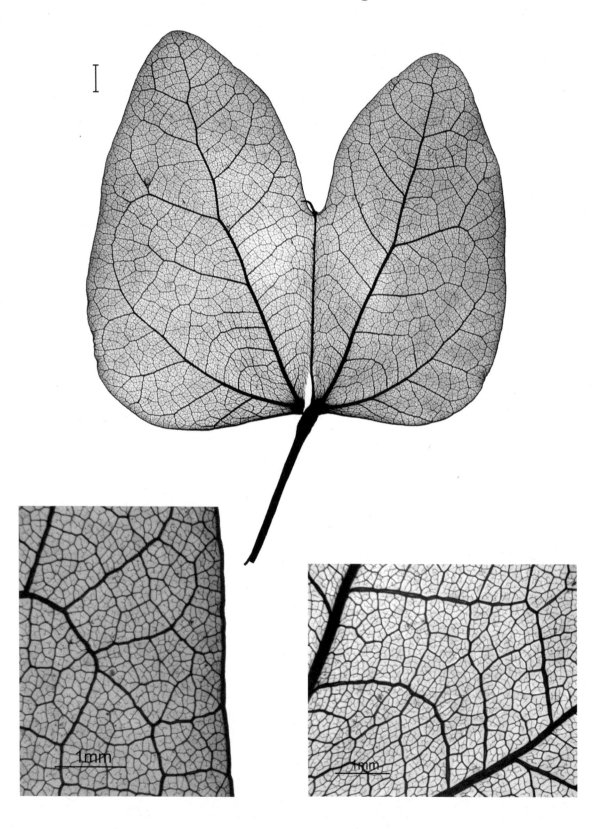

Fabaceae - *Bauhinia madagascariensis*

I. Leaf Characters	Score	Description
Leaf Attachment	1.1	petiolate
Leaf Arrangement	2.1	alternate
Leaf Organization	3.1	simple
Leaflet Arrangement	99	n/a
Leaflet Attachment	99	n/a
Petiole Features	6.1.2	pulvinate

Features of the Blade

Position of Blade Attachment	7.1	marginal
Laminar Size	8.5	mesophyll
Laminar L:W Ratio		0.85:1
Laminar Shape	10.3	ovate
Medial Symmetry	11.2	asymmetrical
Base Symmetry	12.1.1	basal width asymmetrical
Base Symmetry	12.1.1	basal width asymmetrical
Lobation	13.2.4	bilobed
Margin Type	14.1	untoothed
Special Margin Features	88	not visible
Apex Angle	16.3	reflex
Apex Shape	17.5	lobed
Base Angle	18.3	reflex
Base Shape	19.2.1	cordate
Base Shape	19.2.1	cordate
Terminal Apex Features	20.2	spinose
Surface Texture	88	not visible
Surficial Glands	88	not visible

	II. Venation	Score	Description
1°	Primary Vein Framework	23.2.1.1	basal actinodromous
	Naked Basal Veins	24.1	absent
	Number of Basal Veins		5
	Agrophic Veins	26.2.1	simple
2°	Major 2° Vein Framework	27.3.2	festooned brochidodromous
	Interior Secondaries	28.2	present
	Minor Secondary Course	29.2	simple brochidodromous
	Perimarginal Veins	30.3	fimbrial vein
	Major Secondary Spacing	31.5	abruptly increasing proximally
	Variation of Secondary Angle	32.1	uniform
	Major Secondary Attachment	33.4	deflected
Inter-2°	Proximal Course	34.1.1	parallel to major secondaries
	Length	34.2.2	>50% of subjacent secondary
	Distal Course	34.3.1	reticulating
	Vein Frequency	34.4.1	<1 per intercostal area
3°	Intercostal 3° Vein Fabric	35.1.1.3	mixed percurrent
	Angle of Percurrent Tertiaries	35.1.2.2	obtuse
	Vein Angle Variability	36.5	increasing proximally
	Epimedial Tertiaries	37.1.1.1	opposite percurrent
	Admedial Course	37.2.1.6	acute to midvein
	Exmedial Course	37.2.2.2	basiflexed
	Exterior Tertiary Course	38.2	looped
4°	Quaternary Vein Fabric	39.2	regular reticulate
5°	Quinternary Vein Fabric	40.1.1	regular reticulate
	Areolation	41.2.3	good development
	FEV branching	42.1.2	mostly unbranched
	FEV termination	42.2.1	simple
	Marginal Ultimate Venation	43.1	absent

III. Teeth	Score	Description
Tooth Spacing	99	n/a
Number of Orders of Teeth	99	n/a
Teeth / cm	99	n/a
Sinus Shape	99	n/a
Tooth Shapes	99	n/a
Tooth Shapes	99	n/a
Tooth Shapes	99	n/a
Tooth Shapes	99	n/a
Principal Vein	99	n/a
Principal Vein Termination	99	n/a
Course of Accessory Vein	99	n/a
Features of the Tooth Apex	99	n/a

Text Description:

Leaf attachment petiolate. Blade attachment marginal, laminar size mesophyll, L:W ratio 0.85:1, laminar shape ovate with medial asymmetry and basal width asymmetry. Margin is bilobed and untoothed with reflex apex angle, lobed apex shape, spinose apex, reflex base angle, and cordate base shape. Primary venation basal actinodromous with no naked basal veins, five basal veins, and simple agrophic veins. Major secondaries simple brochidodromous with spacing that abruptly increases proximally, uniform angle, and decurrent attachment to midvein. Interior secondaries present, minor secondaries simple brochidodromous, and fimbrial vein present. Intersecondaries span more than 50% of the length of the subjacent secondary, occur at less than one per intercostal area, proximal course is parallel to major secondaries, and distal course is reticulating or basiflexed. Intercostal tertiary veins mixed percurrent to irregular reticulate. Epimedial tertiaries opposite percurrent with proximal course acute to the midvein and distal course basiflexed. Exterior tertiaries looped. Quaternary vein fabric regular reticulate. Quinternary vein fabric irregular reticulate. Areolation shows good development. Freely ending veinlets mostly unbranched, and marginal ultimate venation is absent.

Example 5. Trochodendraceae - *Tetracentron sinense*

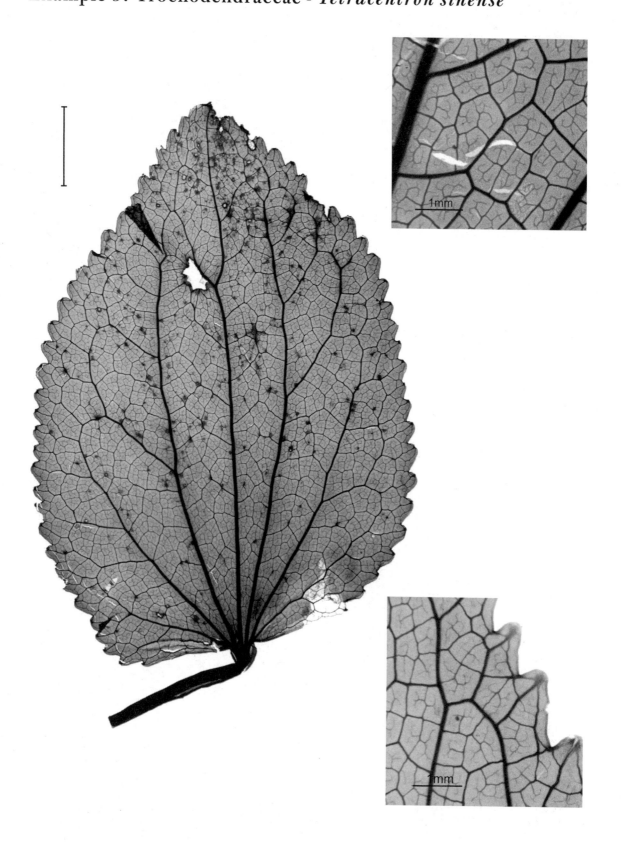

Trochodendraceae - *Tetracentron sinense*

I. Leaf Characters	Score	Description
Leaf Attachment	1.1	petiolate
Leaf Arrangement	2.1	alternate
Leaf Organization	3.1	simple
Leaflet Arrangement	99	n/a
Leaflet Attachment	99	n/a
Petiole Features	88	not visible

Features of the Blade

	Score	Description
Position of Blade Attachment	7.1	marginal
Laminar Size	8.4	notophyll
Laminar L:W Ratio		1.3:1
Laminar Shape	10.3	ovate
Medial Symmetry	11.1	symmetrical
Base Symmetry	12.2.1	basal width asymmetrical
Base Symmetry	12.2.1	basal width asymmetrical
Lobation	13.1	unlobed
Margin Type	14.2.2	serrate
Special Margin Features	0	absent
Apex Angle	16.1	acute
Apex Shape	17.1	straight
Base Angle	18.2	obtuse
Base Angle	18.2	obtuse
Base Shape	19.1.3.2	truncate
Terminal Apex Features	0	absent
Surface Texture	88	not visible
Surficial Glands	88	not visible

	II. Venation	Score	Description
1°	Primary Vein Framework	23.2.1.1	basal actinodromous
	Naked Basal Veins	24.1	absent
	Number of Basal Veins		7
	Agrophic Veins	26.2.2	compound
2°	Major 2° Vein Framework	27.1.3	festooned semicraspedodromous
	Interior Secondaries	28.1	absent
	Minor Secondary Course	29.3	semicraspedodromous
	Perimarginal Veins	0	absent
	Major Secondary Spacing	31.4	gradually increasing proximally
	Variation of Secondary Angle	32.1	uniform
	Major Secondary Attachment	33.4	deflected
Inter-2°	Proximal Course	0	absent
	Length	99	n/a
	Distal Course	99	n/a
	Vein Frequency	99	n/a
3°	Intercostal 3° Vein Fabric	35.2.2	regular reticulate
	Angle of Percurrent Tertiaries	99	n/a
	Vein Angle Variability	99	n/a
	Epimedial Tertiaries	37.1.3	reticulate
	Admedial Course	99	n/a
	Exmedial Course	99	n/a
	Exterior Tertiary Course	38.4	variable
4°	Quaternary Vein Fabric	39.2.1	regular reticulate
5°	Quinternary Vein Fabric	0	absent
	Areolation	41.2.2	moderate development
	FEV branching	42.1.4.2	2 or more, dendritic
	FEV termination	42.2.1	simple
	Marginal Ultimate Venation	43.3	looped

III. Teeth	Score	Description
Tooth Spacing	44.1	regular
Number of Orders of Teeth	45.1	one
Teeth / cm		4
Sinus Shape	47.1	angular
Tooth Shapes		cv/cv
Tooth Shapes		
Tooth Shapes		
Tooth Shapes		
Tooth Shapes		
Principal Vein	49.1	present
Principal Vein Termination	50.2.1	at apex of tooth
Course of Accessory Vein	51.1	convex
Features of the Tooth Apex	52.2.3	cassidate

Text Description:

Leaf attachment petiolate. Blade attachment marginal, laminar size notophyll, L:W ratio 1.3:1, laminar shape ovate with medial symmetry and basal width asymmetry. Margin is unlobed and serrate with acute apex angle, straight apex shape, obtuse base angle, and truncate base shape. Primary venation is basal actinodromous with no naked basal veins, seven basal veins, and compound agrophic veins. Major secondaries festooned semicraspedodromous with spacing that gradually increases proximally, uniform angle, and deflected attachment to midvein. Interior secondaries absent, minor secondaries semicraspedodromous, and perimarginal veins absent. Intersecondaries absent. Intercostal tertiary veins regular reticulate. Epimedial tertiaries reticulate. Exterior tertiaries variable. Quaternary vein fabric regular reticulate. Areolation moderately developed. Freely ending veinlets mostly two or more branched with simple termination. Marginal ultimate venation looped. Tooth spacing regular with one order of teeth and 4 teeth/cm. Sinus shape angular and tooth shape convex/convex. Principal vein present and terminating at tooth apex. Accessory vein course convex. Tooth apex cassidate.

Example 6. Anacardiaceae - *Buchanania arborescens*

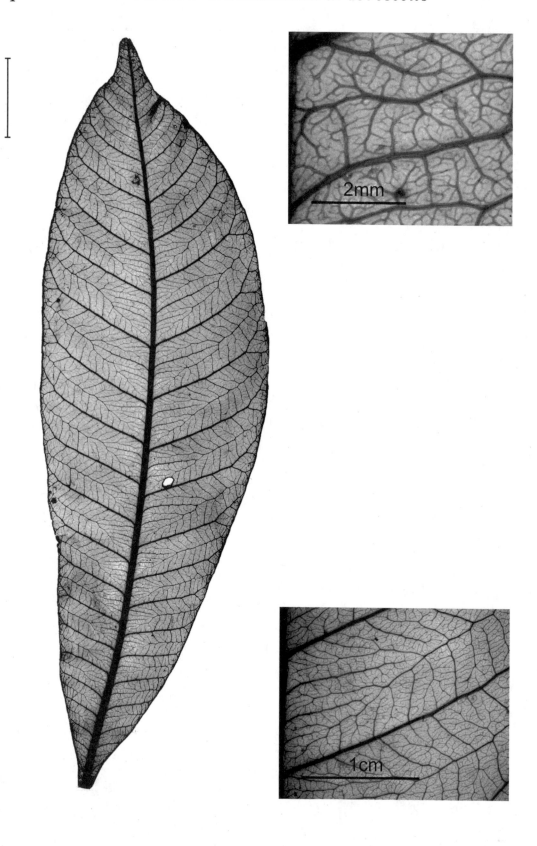

Anacardiaceae - *Buchanania arborescens*

I. Leaf Characters	Score	Description
Leaf Attachment	1.1	petiolate
Leaf Arrangement	2.1	alternate
Leaf Organization	3.1	simple
Leaflet Arrangement	99	n/a
Leaflet Attachment	99	n/a
Petiole Features	88	not visible

Features of the Blade

	Score	Description
Position of Blade Attachment	7.1	marginal
Laminar Size	8.5	mesophyll
Laminar L:W Ratio		3:1
Laminar Shape	10.2	obovate
Medial Symmetry	11.1	symmetrical
Base Symmetry	12.1	symmetrical
Base Symmetry	12.1	symmetrical
Lobation	13.1	unlobed
Margin Type	14.1	untoothed
Special Margin Features	88	not visible
Apex Angle	18.1	acute
Apex Shape	17.3	acuminate
Base Angle	18.1	acute
Base Shape	19.1.2	concave
Base Shape	19.1.2	concave
Terminal Apex Features	0	absent
Surface Texture	88	not visible
Surficial Glands	88	not visible

	II. Venation	Score	Description
1°	Primary Vein Framework	23.1	pinnate
	Naked Basal Veins	24.1	absent
	Number of Basal Veins		1
	Agrophic Veins	26.1	absent
2°	Major 2° Vein Framework	27.2.3	cladodromous
	Interior Secondaries	28.1	absent
	Minor Secondary Course	0	n/a
	Perimarginal Veins	30.3	fimbrial vein
	Major Secondary Spacing	31.3	decreasing proximally
	Variation of Secondary Angle	32.3	smoothly increasing proximally
	Major Secondary Attachment	33.1	decurrent
Inter-2°	Proximal Course	0	absent
	Length	99	n/a
	Distal Course	99	n/a
	Vein Frequency	99	n/a
3°	Intercostal 3° Vein Fabric	35.2.3	composite admedial
	Angle of Percurrent Tertiaries	99	n/a
	Vein Angle Variability	99	n/a
	Epimedial Tertiaries	37.1.2	ramified
	Admedial Course	37.2.1.1	parallel to subjacent secondary
	Exmedial Course		ramified
	Exterior Tertiary Course	38.4	variable
4°	Quaternary Vein Fabric	39.3	freely ramifying
5°	Quinternary Vein Fabric	0	absent
	Areolation	41.2.1	poorly developed
	FEV branching	42.1.4.2	2 or more, dendritic
	FEV termination	42.2.1	simple
	Marginal Ultimate Venation	43.1	absent

III. Teeth	Score	Description
Tooth Spacing	99	n/a
Number of Orders of Teeth	99	n/a
Teeth / cm	99	n/a
Sinus Shape	99	n/a
Tooth Shapes	99	n/a
Tooth Shapes	99	n/a
Tooth Shapes	99	n/a
Tooth Shapes	99	n/a
Principal Vein	99	n/a
Principal Vein Termination	99	n/a
Course of Accessory Vein	99	n/a
Features of the Tooth Apex	99	n/a

Text Description:

Blade attachment marginal, laminar size mesophyll, L:W ratio 3:1, laminar shape obovate with medial symmetry and basal symmetry. Margin entire with acute apex angle, acuminate apex, acute base angle, and concave base shape. Primary venation is pinnate with no naked basal veins, one basal vein, and no agrophic veins. Major secondaries cladodromous with spacing that decreases proximally, angle that smoothly increases proximally, and decurrent attachment to midvein. Interior secondaries absent, minor secondaries absent, and fimbrial vein present. Intersecondaries absent. Intercostal tertiary veins composite admedial. Epimedial tertiaries ramified with admedial course parallel to subjacent secondary and exmedial course ramified. Exterior tertiaries variable. Quaternary vein fabric freely ramifying. Areolation poorly developed. Freely ending veinlets have two or more dendritic branches.

Example 7. Elaeocarpaceae - *Aristotelia racemosa*

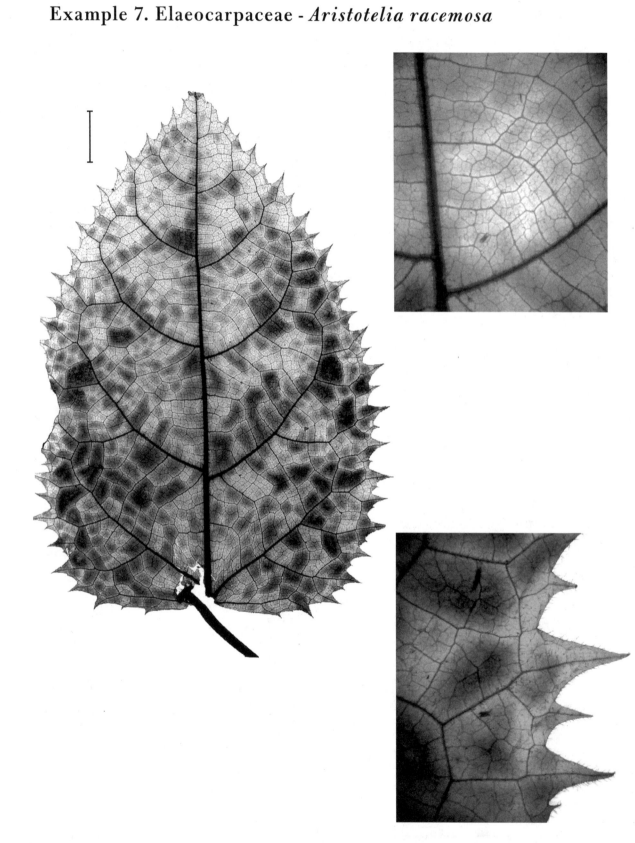

Elaeocarpaceae - *Aristotelia racemosa*

I. Leaf Characters	Score	Description
Leaf Attachment	1.1	petiolate
Leaf Arrangement	2.3	opposite
Leaf Organization	3.1	simple
Leaflet Arrangement	99	n/a
Leaflet Attachment	99	n/a
Petiole Features	88	not visible

Features of the Blade

	Score	Description
Position of Blade Attachment	7.1	marginal
Laminar Size	8.5	mesophyll
Laminar L:W Ratio		1.4:1
Laminar Shape	10.3	ovate
Medial Symmetry	11.1	symmetrical
Base Symmetry	12.1	symmetrical
Base Symmetry	12.1	symmetrical
Lobation	13.1	unlobed
Margin Type	14.2.1	dentate
Special Margin Features	88	not visible
Apex Angle	16.1	acute
Apex Shape	17.1	straight
Base Angle	18.2	obtuse
Base Shape	19.1.3.2	truncate
Base Shape	19.1.3.2	truncate
Terminal Apex Features	0	absent
Surface Texture	88	not visible
Surficial Glands	88	not visible

	II. Venation	Score	Description
1°	Primary Vein Framework	23.1	pinnate
	Naked Basal Veins	24.1	absent
	Number of Basal Veins		5
	Agrophic Veins	26.2.2	compound
2°	Major 2° Vein Framework	27.1.3	festooned semicraspedodromous
	Interior Secondaries	28.1	absent
	Minor Secondary Course	29.3	semicraspedodromous
	Perimarginal Veins	0	absent
	Major Secondary Spacing	31.4	gradually increasing proximally
	Variation of Secondary Angle	32.1	uniform
	Major Secondary Attachment	33.3	excurrent
Inter-2°	Proximal Course	0	absent
	Length	99	n/a
	Distal Course	99	n/a
	Vein Frequency	99	n/a
3°	Intercostal 3° Vein Fabric	35.1.1.3	mixed percurrent
	Angle of Percurrent Tertiaries	35.1.2.2	obtuse
	Vein Angle Variability	36.5	increasing proximally
	Epimedial Tertiaries	37.1.1.3	mixed percurrent
	Admedial Course	37.2.1.3	perpendicular to midvein
	Exmedial Course	37.2.2.1	parallel to intercostal tertiary
	Exterior Tertiary Course	38.3	terminating at the margin
4°	Quaternary Vein Fabric	39.2.1	regular reticulate
5°	Quinternary Vein Fabric	40.1.1	regular reticulate
	Areolation	41.2.3	good development
	FEV branching	42.1.2	mostly unbranched
	FEV termination	42.2.1	simple
	Marginal Ultimate Venation	43.4	looped

III. Teeth	Score	Description
Tooth Spacing	44.1	regular
Number of Orders of Teeth	45.2	two
Teeth / cm		2
Sinus Shape	47.2	rounded
Tooth Shapes		st/st
Tooth Shapes		cc/cc
Tooth Shapes		
Tooth Shapes		
Principal Vein	49.1	present
Principal Vein Termination	50.2.1	at apex of tooth
Course of Accessory Vein	51.2	straight or concave
Features of the Tooth Apex	52.1	simple

Text Description:

Leaf attachment petiolate. Blade attachment marginal, laminar size mesophyll, L:W ratio 1.4:1, laminar shape elliptic to ovate with medial symmetry and basal symmetry. Margin is unlobed and dentate with acute apex angle, straight apex shape, obtuse base angle, and truncate base shape. Primary venation is pinnate with no naked basal veins, five basal veins, and compound agrophic veins. Major secondaries festooned semicraspedodromous with spacing that gradually increases proximally, with uniform angle and excurrent attachment to midvein. Interior secondaries absent, minor secondaries semicraspedodromous, and perimarginal veins absent. Intersecondaries absent. Intercostal tertiary veins mixed percurrent with obtuse angle to midvein and proximally increasing vein angle. Epimedial tertiaries mixed percurrent with proximal course perpendicular to the midvein and distal course parallel to intercostal tertiary. Exterior tertiaries terminate at the margin. Quaternary and quinternary vein fabric regular reticulate. Areolation shows good development. Tooth spacing regular with two orders of teeth and 2 teeth/cm. Sinus shape rounded and tooth shape straight/straight to concave/concave. Principal vein present and terminating at tooth apex. Accessory vein course straight or concave. Tooth apex simple.

Example 8. Malvaceae - *Bombacopsis rupicola*

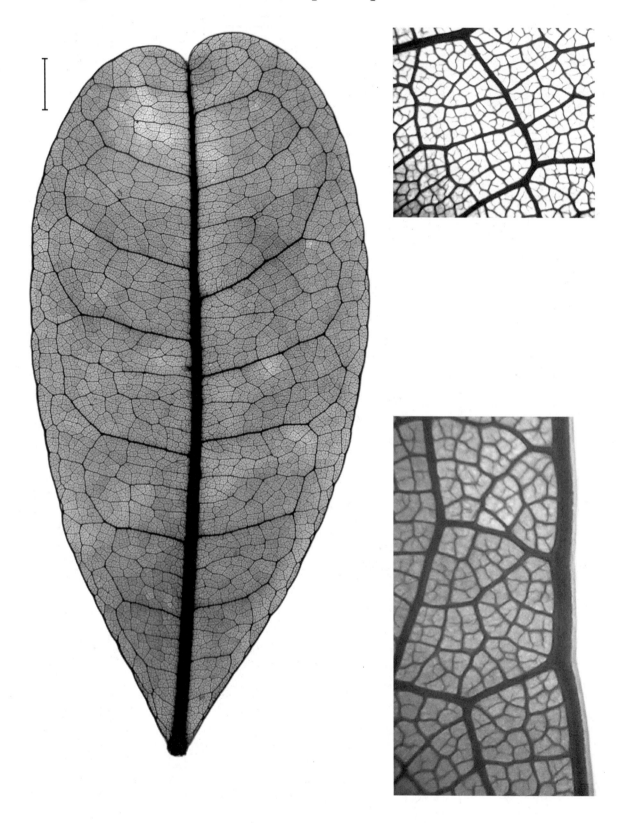

Malvaceae - *Bombacopsis rupicola*

I. Leaf Characters	Score	Description
Leaf Attachment	1.1	petiolate
Leaf Arrangement	2.1	alternate
Leaf Organization	3.2.1	palmately compound
Leaflet Arrangement	99	n/a
Leaflet Attachment	5.1	petiolulate
Petiole Features	88	not visible

Features of the Blade

	Score	Description
Position of Blade Attachment	7.1	marginal
Laminar Size	8.5	mesophyll
Laminar L:W Ratio		2.2:1
Laminar Shape	10.2	obovate
Medial Symmetry	11.1	symmetrical
Base Symmetry	12.1	symmetrical
Base Symmetry	12.1	symmetrical
Lobation	13.1	unlobed
Margin Type	14.1	untoothed
Special Margin Features	88	not visible
Apex Angle	16.3	reflex
Apex Shape	17.2	convex
Base Angle	18.1	acute
Base Shape	19.1.1	straight
Base Shape	19.1.1	straight
Terminal Apex Features	20.3	retuse
Surface Texture	88	not visible
Surficial Glands	88	not visible

	II. Venation	Score	Description
1°	Primary Vein Framework	23.1	pinnate
	Naked Basal Veins	24.1	absent
	Number of Basal Veins		1
	Agrophic Veins	26.1	absent
2°	Major 2° Vein Framework	27.3.2	festooned brochidodromous
	Interior Secondaries	28.1	absent
	Minor Secondary Course	0	absent
	Perimarginal Veins	30.1	marginal secondary
	Major Secondary Spacing	31.3	decreasing proximally
	Variation of Secondary Angle	32.2	inconsistent
	Major Secondary Attachment	33.3	excurrent
Inter-2°	Proximal Course	34.1.3	perpendicular to midvein
	Length	34.2.2	>50% of subjacent secondary
	Distal Course	34.3.1	reticulating
	Vein Frequency	34.4.2	~1 per intercostal area
3°	Intercostal 3° Vein Fabric	35.2.1	irregular reticulate
	Angle of Percurrent Tertiaries	99	n/a
	Vein Angle Variability	99	n/a
	Epimedial Tertiaries	37.1.3	reticulate
	Admedial Course	99	n/a
	Exmedial Course	99	n/a
	Exterior Tertiary Course	38.2	looped
4°	Quaternary Vein Fabric	39.2.2	irregular reticulate
5°	Quinternary Vein Fabric	40.1.1	regular reticulate
	Areolation	41.2.3	good development
	FEV branching	42.1.2	mostly 1 branch
	FEV termination	42.2.1	simple
	Marginal Ultimate Venation	0	n/a

III. Teeth	Score	Description
Tooth Spacing	99	n/a
Number of Orders of Teeth	99	n/a
Teeth / cm	99	n/a
Sinus Shape	99	n/a
Tooth Shapes	99	n/a
Tooth Shapes	99	n/a
Tooth Shapes	99	n/a
Tooth Shapes	99	n/a
Principal Vein	99	n/a
Principal Vein Termination	99	n/a
Course of Accessory Vein	99	n/a
Features of the Tooth Apex	99	n/a

Text Description:

Blade attachment marginal, laminar size notophyll to mesophyll, L:W ratio 2.2:1, laminar shape obovate to elliptic with medial symmetry and basal symmetry. Margin entire with reflex apex angle, convex apex shape and retuse apex, acute base angle, and straight base shape. Primary venation pinnate with no naked basal veins, one basal vein, and no agrophic veins. Major secondaries festooned brochidodromous with spacing that decreases proximally, inconsistent secondary angle, and excurrent attachment to midvein. Minor secondaries absent and marginal secondary present. Intersecondaries span more than 50% of the length of the subjacent secondary, occur at roughly one per intercostal area, proximal course is perpendicular to midvein and distal course is reticulating. Intercostal tertiary veins irregular reticulate. Epimedial tertiaries reticulate. Exterior tertiaries looped. Quaternary vein fabric irregular reticulate. Quinternary vein fabric regular reticulate. Areolation shows good development.

Example 9. Gesneriaceae - *Rhynchoglossum azureum*

Gesneriaceae - *Rhynchoglossum azureum*

I. Leaf Characters	Score	Description
Leaf Attachment	1.1	petiolate
Leaf Arrangement	2.1	alternate
Leaf Organization	3.1	simple
Leaflet Arrangement	99	n/a
Leaflet Attachment	99	n/a
Petiole Features	88	not visible

Features of the Blade

	Score	Description
Position of Blade Attachment	7.1	marginal
Laminar Size	8.4	notophyll
Laminar L:W Ratio		1.85:1
Laminar Shape	10.1	elliptic
Medial Symmetry	11.2	asymmetrical
Base Symmetry	12.2.1	basal width asymmetrical
Base Symmetry	12.2.3	basal insertion asymmetrical
Lobation	13.1	unlobed
Margin Type	14.1	untoothed
Special Margin Features		n/a
Apex Angle	16.1	acute
Apex Shape	17.3	acuminate
Base Angle	18.2	obtuse
Base Shape	19.1.2	concave
Base Shape	19.1.3	convex
Terminal Apex Features	0	n/a
Surface Texture	21.5	pubescent
Surficial Glands	22.2	marginal

II. Venation		Score	Description
1°	Primary Vein Framework	23.1	pinnate
	Naked Basal Veins	24.1	absent
	Number of Basal Veins		1
	Agrophic Veins	26.1	absent
2°	Major 2° Vein Framework	27.2.1	eucamptodromous
	Interior Secondaries	28.1	absent
	Minor Secondary Course	0	absent
	Perimarginal Veins	0	absent
	Major Secondary Spacing	31.3	decreasing proximally
	Variation of Secondary Angle	32.3	smoothly increasing proximally
	Major Secondary Attachment	33.3	excurrent
Inter-2°	Proximal Course	0	absent
	Length	99	n/a
	Distal Course	99	n/a
	Vein Frequency	99	n/a
3°	Intercostal 3° Vein Fabric	35.1.1.1.4	opposite percurrent
	Angle of Percurrent Tertiaries	35.1.2.2	obtuse
	Vein Angle Variability	36.4	decreasing exmedially
	Epimedial Tertiaries	37.1.1.1	opposite percurrent
	Admedial Course	37.2.1.6	acute to midvein
	Exmedial Course	37.2.2.2	basiflexed
	Exterior Tertiary Course	38.2	looped
4°	Quaternary Vein Fabric	39.2.2	irregular reticulate
5°	Quinternary Vein Fabric	40.1.2	irregular reticulate
	Areolation	41.2.1	poor development
	FEV branching	42.1.3	mostly 1 branch
	FEV termination	42.2.1	simple
	Marginal Ultimate Venation	43.4	looped

III. Teeth	Score	Description
Tooth Spacing	99	n/a
Number of Orders of Teeth	99	n/a
Teeth / cm	99	n/a
Sinus Shape	99	n/a
Tooth Shapes	99	n/a
Tooth Shapes	99	n/a
Tooth Shapes	99	n/a
Tooth Shapes	99	n/a
Principal Vein	99	n/a
Principal Vein Termination	99	n/a
Course of Accessory Vein	99	n/a
Features of the Tooth Apex	99	n/a

Text Description:

Blade attachment marginal, laminar size notophyll, L:W ratio 1.85:1, laminar shape elliptic with medial asymmetry and basal width and basal insertion asymmetry. Margin is entire with acute apex angle, acuminate apex shape, obtuse base angle, and concave to rounded base shape. Surface texture is pubescent with surface glands. Primary venation is pinnate with one basal vein, and no agrophic veins. Major secondaries eucamptodromous with spacing that decreases exmedially, mostly attaching to the midvein excurrently but with some apical deflection. Minor secondaries and intersecondaries absent. Intercostal 3° veins form chevrons with vein angles that decrease exmedially. Epimedial tertiaries opposite percurrent with proximal course acute to midvein and distal course basiflexed. Exterior tertiaries looped. Quaternary vein fabric irregular reticulate. Quinternary vein fabric irregular reticulate. Areolation shows poor development. FEV's are mostly one branched with simple terminals. Marginal ulitmate venation looped.

Example 10. Nothofagaceae - *Nothofagus procera*

Nothofagaceae - *Nothofagus procera*

I. Leaf Characters	Score	Description
Leaf Attachment	1.1	petiolate
Leaf Arrangement	2.1	alternate
Leaf Organization	3.1	simple
Leaflet Arrangement	99	n/a
Leaflet Attachment	99	n/a
Petiole Features	88	not visible

Features of the Blade

	Score	Description
Position of Blade Attachment	7.1	marginal
Laminar Size	8.4	notophyll
Laminar L:W Ratio		2.4:1
Laminar Shape	10.1	elliptic
Medial Symmetry	11.1	symmetrical
Base Symmetry	12.1	symmetrical
Base Symmetry	12.1	symmetrical
Lobation	13.1	unlobed
Margin Type	14.2.2	serrate
Special Margin Features	15.1.2	sinuous
Apex Angle	16.1	acute
Apex Shape	17.2	convex
Base Angle	18.2	obtuse
Base Shape	19.1.3	convex
Base Shape	19.1.3	convex
Terminal Apex Features	0	absent
Surface Texture	88	not visible
Surficial Glands	88	not visible

	II. Venation	Score	Description
1°	Primary Vein Framework	23.1	pinnate
	Naked Basal Veins	24.1	absent
	Number of Basal Veins		3
	Agrophic Veins	26.1	absent
2°	Major 2° Vein Framework	27.1.2	semicraspedodromous
	Interior Secondaries	28.1	absent
	Minor Secondary Course	0	absent
	Perimarginal Veins	0	absent
	Major Secondary Spacing	31.1	regular
	Variation of Secondary Angle	32.1	uniform
	Major Secondary Attachment	33.2	basally decurrent
Inter-2°	Proximal Course	0	absent
	Length	99	n/a
	Distal Course	99	n/a
	Vein Frequency	99	n/a
3°	Intercostal 3° Vein Fabric	35.1.1.2	alternate percurrent
	Angle of Percurrent Tertiaries	35.1.2.2	obtuse
	Vein Angle Variability	36.2	consistent
	Epimedial Tertiaries	37.1.3	reticulate
	Admedial Course	99	n/a
	Exmedial Course	99	n/a
	Exterior Tertiary Course	38.3	terminating at the margin
4°	Quaternary Vein Fabric	39.2.1	regular reticulate
5°	Quinternary Vein Fabric	40.1.1	regular reticulate
	Areolation	41.2.3	good development
	FEV branching	42.1.2	mostly unbranched
	FEV termination	42.2.1	simple
	Marginal Ultimate Venation	43.4	looped

III. Teeth	Score	Description
Tooth Spacing	44.2	irregular
Number of Orders of Teeth	45.2	two
Teeth / cm		6
Sinus Shape	47.2	rounded
Tooth Shapes		st/st
Tooth Shapes		cv/cv
Tooth Shapes		
Tooth Shapes		
Principal Vein	49.1	present
Principal Vein Termination	50.2.1	at apex of tooth
Course of Accessory Vein	51.2	straight
Features of the Tooth Apex	52.1	simple

Text Description:

Blade attachment marginal, laminar size microphyll to notophyll, L:W ratio 2.4:1, laminar shape elliptic with medial symmetry and basal symmetry. Margin unlobed, sinuous and serrate, with acute apex angle, convex apex shape, obtuse base angle, and convex base shape. Primary venation pinnate with no naked basal veins, three basal veins, and no agrophic veins. Major secondaries semicraspedodromous with regular spacing, uniform angle, and basally decurrent attachment to midvein. Interior secondaries absent, minor secondaries absent, and perimarginal vein absent. Intersecondaries absent. Intercostal tertiary veins alternate percurrent with obtuse angle to midvein and consistent vein angle. Epimedial tertiaries reticulate. Exterior tertiaries terminate at the margin. Quaternary and quinternary vein fabric regular reticulate. Areolation shows good development. Tooth spacing irregular with two orders of teeth and 6 teeth/cm. Sinus shape rounded and tooth shape straight/straight to convex/convex. Principal vein terminating at tooth apex.

Example 11. Sapindaceae - *Acer franchetii*

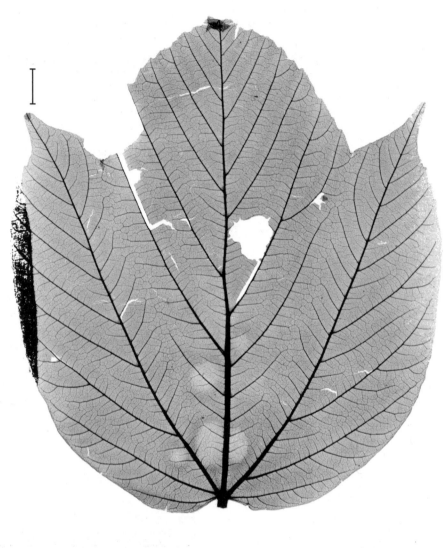

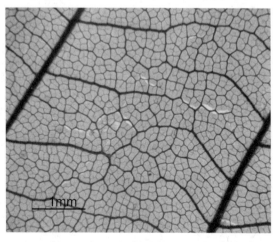

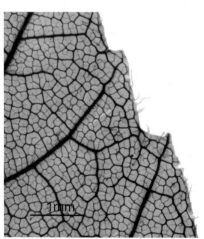

Sapindaceae - *Acer franchetii*

I. Leaf Characters	Score	Description
Leaf Attachment	1.1	petiolate
Leaf Arrangement	2.3	opposite
Leaf Organization	3.1	simple
Leaflet Arrangement	99	n/a
Leaflet Attachment	99	n/a
Petiole Features	88	not visible

Features of the Blade

	Score	Description
Position of Blade Attachment	7.1	marginal
Laminar Size	8.6	macrophyll
Laminar L:W Ratio		1.1:1
Laminar Shape	10.1	elliptic
Medial Symmetry	11.1	symmetrical
Base Symmetry	12.1	symmetrical
Base Symmetry	12.1	symmetrical
Lobation	13.2.1	palmately lobed
Margin Type	14.2.2	serrate
Special Margin Features	0	absent
Apex Angle	16.2	obtuse
Apex Shape	17.2	convex
Base Angle	18.3	reflex
Base Shape	19.2.1	cordate
Base Shape	19.2.1	cordate
Terminal Apex Features	0	absent
Surface Texture	88	not visible
Surficial Glands	88	not visible

	II. Venation	Score	Description
1°	Primary Vein Framework	23.2.1.1	basal actinodromous
	Naked Basal Veins	24.1	absent
	Number of Basal Veins		6
	Agrophic Veins	26.2.2	compound
2°	Major 2° Vein Framework	27.1.1	craspedodromous
	Interior Secondaries	28.2	present
	Minor Secondary Course	29.1	craspedodromous
	Perimarginal Veins	0	absent
	Major Secondary Spacing	31.5	abruptly increasing proximally
	Variation of Secondary Angle	32.1	uniform
	Major Secondary Attachment	33.4	deflected
Inter-2°	Proximal Course	34.1.1	parallel to major secondaries
	Length	34.3.2	parallel to major secondary
	Distal Course	34.2.1	<50% of subjacent secondary
	Vein Frequency	34.4.3	>1 per intercostal area
3°	Intercostal 3° Vein Fabric	35.1.1.2	alternate percurrent
	Angle of Percurrent Tertiaries	35.1.2.2	obtuse
	Vein Angle Variability	36.2	consistent
	Epimedial Tertiaries	37.1.2	ramified
	Admedial Course	37.2.1.6	acute to midvein
	Exmedial Course	37.2.2.1	parallel to intercostal tertiary
	Exterior Tertiary Course	38.2	looped
4°	Quaternary Vein Fabric	39.2.1	regular reticulate
5°	Quinternary Vein Fabric	40.1.1	regular reticulate
	Areolation	41.2.2	moderate development
	FEV branching	42.1.2	mostly unbranched
	FEV termination	42.2.1	simple
	Marginal Ultimate Venation	43.3	spiked

III. Teeth	Score	Description
Tooth Spacing	44.2	irregular
Number of Orders of Teeth	45.1	one
Teeth / cm		3
Sinus Shape	47.1	angular
Tooth Shapes		st/st
Tooth Shapes		cv/cv
Tooth Shapes		
Tooth Shapes		
Principal Vein	49.1	present
Principal Vein Termination	50.2.1	at apex of tooth
Course of Accessory Vein	51.3	running from sinus
Features of the Tooth Apex	52.1	simple

Text Description:

Blade attachment marginal, laminar size microphyll to macrophyll, laminar L:W ratio 1.1:1, laminar shape elliptic, blade medially symmetrical, base symmetrical, palmately lobed, margin serrate. Apex angle obtuse, apex shape convex, base angle reflex, base shape cordate. Primary vein basal actinodromous, naked basal veins absent, six basal veins, agrophic veins compound, major 2° veins craspedodromous, minor secondary course craspedodromous, interior secondaries present, major secondary spacing abruptly increasing proximally, secondary angle uniform, major secondary attachment deflected. Intersecondary length <50% of subjacent secondary, distal course parallel to subjacent major secondary, vein frequency >1 per intercostal area, intercostal tertiary vein fabric opposite percurrent. Epimedial tertiaries ramified, admedial course acute to midvein, exmedial course parallel to intercostal tertiary. Exterior tertiary course looped and occasionally terminating at the margin. Quaternary vein fabric regular reticulate; quinternary vein fabric regular reticulate; areolation development moderate. Tooth spacing irregular, one order of teeth, 3 teeth/cm, sinus shape angular, tooth shapes st/st and cv/cv. Principal vein terminates at apex of tooth, accessory veins run from sinus.

Example 12. Malpighiaceae - *Tetrapterys macrocarpa*

Malpighiaceae - *Tetrapterys macrocarpa*

I. Leaf Characters	Score	Description
Leaf Attachment	1.1	petiolate
Leaf Arrangement	2.3	opposite
Leaf Organization	3.1	simple
Leaflet Arrangement	0	n/a
Leaflet Attachment	0	n/a
Petiole Features	6.2.1	glands petiolar

Features of the Blade

	Score	Description
Position of Blade Attachment	7.1	marginal
Laminar Size	8.4	notophyll
Laminar L:W Ratio		1.5:1
Laminar Shape	10.1	elliptic
Medial Symmetry	11.1	symmetrical
Base Symmetry	12.1	symmetrical
Base Symmetry	12.1	symmetrical
Lobation	13.1	unlobed
Margin Type	14.1	untoothed
Special Margin Features	0	absent
Apex Angle	16.2	obtuse
Apex Shape	17.3	acuminate
Base Angle	18.3	reflex
Base Shape	19.2.1	cordate
Base Shape	19.2.1	cordate
Terminal Apex Features	0	absent
Surface Texture	88	not visible
Surficial Glands	88	not visible

	II. Venation	Score	Description
1°	Primary Vein Framework	23.1	pinnate
	Naked Basal Veins	24.1	absent
	Number of Basal Veins		3
	Agrophic Veins	26.1	absent
2°	Major 2° Vein Framework	27.3.2	festooned brochidodromous
	Interior Secondaries	28.1	absent
	Minor Secondary Course	0	absent
	Perimarginal Veins	30.3	fimbrial vein
	Major Secondary Spacing	31.3	decreasing proximally
	Variation of Secondary Angle	32.4	smoothly decreasing proximally
	Major Secondary Attachment	33.1	decurrent
Inter-2°	Proximal Course	34.1.1	parallel to major secondaries
	Length	34.2.1	<50% of subjacent secondary
	Distal Course	34.3.1	reticulating
	Vein Frequency	34.4.1	<1 per intercostal area
3°	Intercostal 3° Vein Fabric	35.1.1.3	mixed percurrent
	Angle of Percurrent Tertiaries	35.1.2.2	obtuse
	Vein Angle Variability	36.1	inconsistent
	Epimedial Tertiaries	37.1.1.1	mixed percurrent
	Admedial Course	37.2.1.3	perpendicular to midvein
	Exmedial Course	37.2.2.1	parallel to intercostal tertiary
	Exterior Tertiary Course	38.2	looped
4°	Quaternary Vein Fabric	39.2.2	irregular reticulate
5°	Quinternary Vein Fabric	40.2.2	irregular reticulate
	Areolation	41.2.2	moderate development
	FEV branching	42.1.4.2	2 or more, dendritic
	FEV termination	42.2.3	highly branched sclereids
	Marginal Ultimate Venation	43.4	looped

III. Teeth	Score	Description
Tooth Spacing	99	n/a
Number of Orders of Teeth	99	n/a
Teeth / cm	99	n/a
Sinus Shape	99	n/a
Tooth Shapes	99	n/a
Tooth Shapes	99	n/a
Tooth Shapes	99	n/a
Tooth Shapes	99	n/a
Principal Vein	99	n/a
Principal Vein Termination	99	n/a
Course of Accessory Vein	99	n/a
Features of the Tooth Apex	99	n/a

Text Description:

Blade attachment marginal, laminar size notophyll, L:W ratio 1.5:1, laminar shape elliptic with medial symmetry and basal symmetry. Margin entire with obtuse apex angle, acuminate apex shape, reflex base angle, and cordate base shape. Primary venation pinnate with no naked basal veins, three basal veins, and no agrophic veins. Major secondaries festooned brochidodromous with spacing that decreases proximally, uniform secondary angle, and decurrent attachment to midvein. Minor secondaries absent, interior secondaries absent. Intersecondaries span less than 50% of the length of the subjacent secondary, occur at less than one per intercostal area, proximal course is parallel to major secondary and distal course is reticulating. Intercostal tertiary veins mixed percurrent with obtuse angle to midvein and inconsistent vein angle variability. Epimedial tertiaries mixed percurrent with proximal course perpendicular to the midvein and distal course parallel to intercostal tertiaries. Exterior tertiaries looped. Quaternary vein fabric irregular reticulate. Quinternary vein fabric irregular reticulate. Areolation shows moderate development. Freely ending veinlets are two or more branched with highly branched sclereids. Marginal ultimate venation forms incomplete loops.

Example 13. Cunoniaceae - *Eucryphia glutinosa*

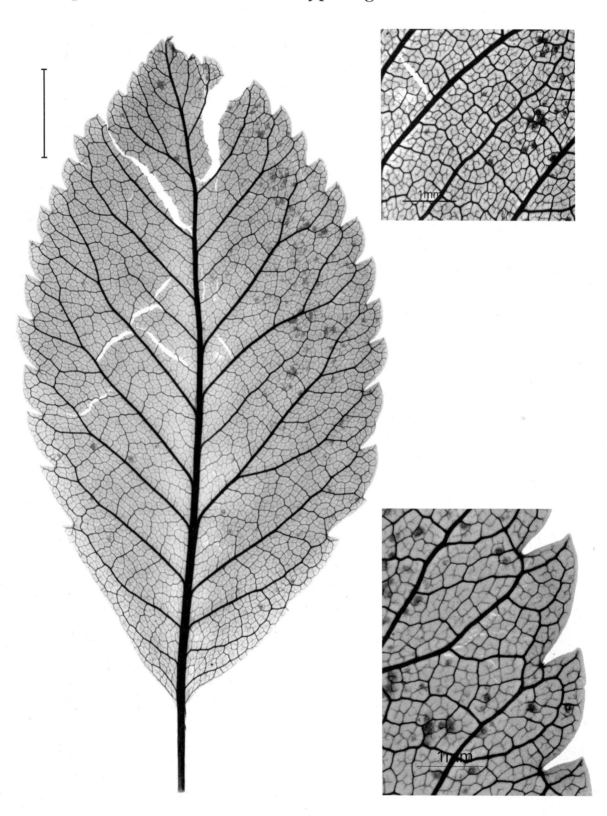

Cunoniaceae - *Eucryphia glutinosa*

I. Leaf Characters	Score	Description
Leaf Attachment	1.1	n/a
Leaf Arrangement	2.3	opposite
Leaf Organization	3.2.2.1	pinnately compounded
Leaflet Arrangement	4.3.2	opposite-even
Leaflet Attachment	5.1	petiolulate
Petiole Features	88	not visible

Features of the Blade

	Score	Description
Position of Blade Attachment	7.1	marginal
Laminar Size	8.4	notophyll
Laminar L:W Ratio		1.8:1
Laminar Shape	10.1	elliptic
Medial Symmetry	11.1	symmetrical
Base Symmetry	12.1	symmetrical
Base Symmetry	12.1	symmetrical
Lobation	13.1	unlobed
Margin Type	14.2.2	serrate
Special Margin Features	0	absent
Apex Angle	16.2	obtuse
Apex Shape	17.2	convex
Base Angle	18.1	acute
Base Shape	19.1.1	straight
Base Shape	19.1.1	straight
Terminal Apex Features	0	absent
Surface Texture	88	not visible
Surficial Glands	88	not visible

	II. Venation	Score	Description
1°	Primary Vein Framework	23.1	pinnate
	Naked Basal Veins	24.1	absent
	Number of Basal Veins		1
	Agrophic Veins	26.1	absent
2°	Major 2° Vein Framework	27.1.2	semicraspedodromous
	Interior Secondaries	28.1	absent
	Minor Secondary Course	0	absent
	Perimarginal Veins	0	absent
	Major Secondary Spacing	31.2	irregular
	Variation of Secondary Angle	32.4	smoothly decreasing proximally
	Major Secondary Attachment	33.1	decurrent
Inter-2°	Proximal Course	34.1.1	parallel to major secondaries
	Length	34.2.2	>50% of subjacent secondary
	Distal Course	34.3.2	parallel to subjacent major 2°
	Vein Frequency	34.4.1	<1 per intercostal area
3°	Intercostal 3° Vein Fabric	35.2.1	irregular reticulate
	Angle of Percurrent Tertiaries	99	n/a
	Vein Angle Variability	99	n/a
	Epimedial Tertiaries	37.1.3	reticulate
	Admedial Course	99	n/a
	Exmedial Course	99	n/a
	Exterior Tertiary Course	38.2	looped
4°	Quaternary Vein Fabric	39.2.2	irregular reticulate
5°	Quinternary Vein Fabric	40.2.2	irregular reticulate
	Areolation	41.2.2	moderate development
	FEV branching	42.1.2	mostly unbranched
	FEV termination	42.2.1	simple
	Marginal Ultimate Venation	43.2	incomplete loops

III. Teeth	Score	Description
Tooth Spacing	44.1	regular
Number of Orders of Teeth	45.1	one
Teeth / cm	3	
Sinus Shape	47.1	angular
Tooth Shapes		cv/cv
Tooth Shapes		st/cv
Tooth Shapes		
Tooth Shapes		
Principal Vein	49.1	present
Principal Vein Termination	50.1	submarginal
Course of Accessory Vein	51.1.1	looped
Features of the Tooth Apex	52.1	simple

Text Description:

Blade attachment marginal, laminar size notophyll, laminar L:W ratio 1.8:1, laminar shape elliptic, blade medially symmetrical, base medially symmetrical, margin unlobed with serrate teeth. Apex angle obtuse, apex shape convex, base angle acute, base shape straight. Primary venation pinnate with one basal vein and no agrophic veins. Secondary veins semicraspedodromous with no interior secondaries, minor secondaries or perimarginal veins. Major secondary spacing smoothly decreasing proximally, major secondary attachment decurrent. Intersecondary proximal course parallel to major secondaries, length >50% of subjacent secondary, distal course parallel to subjacent secondary and frequency less than one per intercostal area. Tertiary vein fabric irregular reticulate with reticulate epimedial tertiaries and looped exterior tertiaries. Quaternary vein fabric irregular reticulate. Quinternary vein fabric irregular reticulate. Areolation moderately developed and FEVs mostly branched with simple terminals. Marginal ultimate venation forms incomplete loops. Tooth spacing is regular with one order of teeth and three teeth per cm. Sinus shape is angular, tooth shapes are convex/convex to straight/convex. Principal vein is present with submarginal termination, looped accessory veins and a simple tooth apex.

Example 14. Chrysobalanaceae - *Licania michauxii*

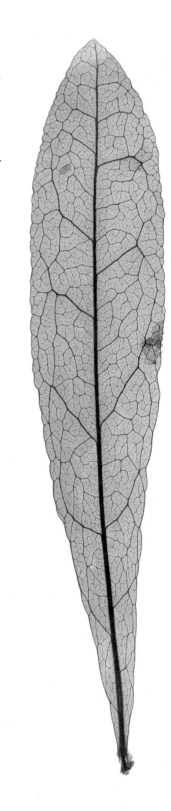

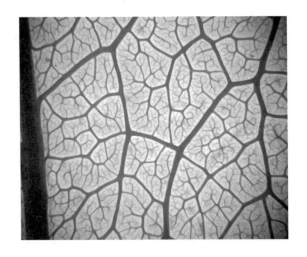

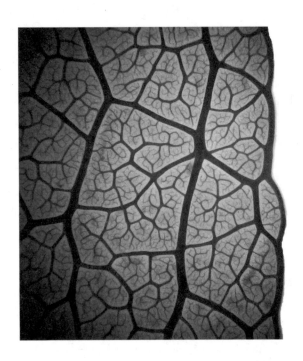

Chrysobalanaceae - *Licania michauxii*

I. Leaf Characters	Score	Description
Leaf Attachment	1.1	petiolate
Leaf Arrangement	2.1	alternate
Leaf Organization	3.1	simple
Leaflet Arrangement	99	n/a
Leaflet Attachment	99	n/a
Petiole Features	88	not visible

Features of the Blade

	Score	Description
Position of Blade Attachment	7.1	marginal
Laminar Size	8.4	microphyll
Laminar L:W Ratio		5:1
Laminar Shape	10.2	obovate
Medial Symmetry	11.1	symmetrical
Base Symmetry	12.1	symmetrical
Base Symmetry	12.1	symmetrical
Lobation	13.1	unlobed
Margin Type	14.1	untoothed
Special Margin Features	15.1.1	erose
Apex Angle	16.1	acute
Apex Shape	17.2	convex
Base Angle	18.1	acute
Base Shape	19.1.1	straight
Base Shape		decurrent
Terminal Apex Features	0	absent
Surface Texture	88	not visible
Surficial Glands	88	not visible

		II. Venation	Score	Description
1°		Primary Vein Framework	23.1	pinnate
		Naked Basal Veins	24.1	absent
		Number of Basal Veins		1
		Agrophic Veins	26.1	absent
2°		Major 2° Vein Framework	27.3.1	simple brochidodromous
		Interior Secondaries	28.1	absent
		Minor Secondary Course	0	absent
		Perimarginal Veins	30.3	fimbrial vein
		Major Secondary Spacing	31.2	irregular
		Variation of Secondary Angle	32.2	inconsistent
		Major Secondary Attachment	33.1	decurrent
Inter-2°		Proximal Course	0	absent
		Length	99	n/a
		Distal Course	99	n/a
		Vein Frequency	99	n/a
3°		Intercostal 3° Vein Fabric	35.2.1	irregular reticulate
		Angle of Percurrent Tertiaries	99	n/a
		Vein Angle Variability	99	n/a
		Epimedial Tertiaries	37.1.3	reticulate
		Admedial Course	99	n/a
		Exmedial Course	99	n/a
		Exterior Tertiary Course	38.3	terminates at the margin
4°		Quaternary Vein Fabric	39.2.2	irregular reticulate
5°		Quinternary Vein Fabric	40.3	freely ramifying
		Areolation	41.2.2	moderate development
		FEV branching	42.1.3	mostly 1 branched
		FEV termination	42.2.1	simple
		Marginal Ultimate Venation	43.1	absent

III. Teeth	Score	Description
Tooth Spacing	99	n/a
Number of Orders of Teeth	99	n/a
Teeth / cm	99	n/a
Sinus Shape	99	n/a
Tooth Shapes	99	n/a
Tooth Shapes	99	n/a
Tooth Shapes	99	n/a
Tooth Shapes	99	n/a
Principal Vein	99	n/a
Principal Vein Termination	99	n/a
Course of Accessory Vein	99	n/a
Features of the Tooth Apex	99	n/a

Text Description:

Blade attachment marginal, laminar size microphyll to notophyll, L:W ratio 5:1, laminar shape obovate with medial symmetry and basal symmetry. Margin is entire and erose with acute apex angle, convex apex shape, acute base angle, and straight to decurrent base shape. Primary venation is pinnate with no naked basal veins, one basal vein, and no agrophic veins. Major secondaries simple brochidodromous with irregular spacing, inconsistent secondary angle, and decurrent attachment to midvein. Minor secondaries, interior secondaries, and intersecondaries absent. Fimbrial vein present. Intercostal tertiary veins irregular reticulate. Epimedial tertiaries reticulate. Exterior tertiaries terminate at the margin. Quaternary vein fabric irregular reticulate. Quinternary vein fabric freely ramifying. Areolation shows moderate development. Freely ending veinlets are mostly one-branched with simple termination. Marginal ultimate venation is absent.

Example 15. Moraceae - *Morus microphylla*

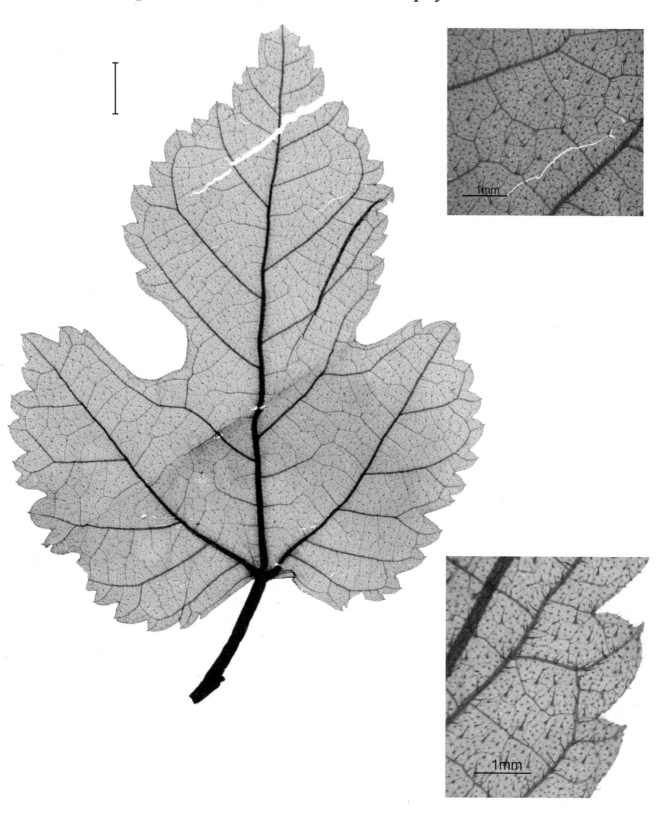

Moraceae - *Morus microphylla*

I. Leaf Characters	Score	Description
Leaf Attachment	1.1	petiolate
Leaf Arrangement	2.1	alternate
Leaf Organization	3.1	simple
Leaflet Arrangement	99	n/a
Leaflet Attachment	99	n/a
Petiole Features	88	not visible

Features of the Blade

	Score	Description
Position of Blade Attachment	7.1	marginal
Laminar Size	8.3	microphyll
Laminar L:W Ratio		1.3:1
Laminar Shape	10.3	ovate
Medial Symmetry	11.2	asymmetrical
Base Symmetry	12.2.1	basal width asymmetrical
Base Symmetry	12.2.1	basal width asymmetrical
Lobation	13.2.1	palmately lobed
Margin Type	14.2.2	serrate
Special Margin Features	88	not visible
Apex Angle	16.1	acute
Apex Shape	17.3	acuminate
Base Angle	18.3	reflex
Base Shape	19.2.1	cordate
Base Shape		
Terminal Apex Features	0	absent
Surface Texture	21.5	pubescent
Surficial Glands	88	not visible

		II. Venation	Score	Description
1°		Primary Vein Framework	23.2.1.1	basal actinodromous
		Naked Basal Veins	24.1	absent
		Number of Basal Veins		5
		Agrophic Veins	26.2.1	simple
2°		Major 2° Vein Framework	27.1.2	semicraspedodromous
		Interior Secondaries	28.1	absent
		Minor Secondary Course	29.3	semicraspedodromous
		Perimarginal Veins	0	absent
		Major Secondary Spacing	31.2	irregular
		Variation of Secondary Angle	32.4	smoothly decreasing proximally
		Major Secondary Attachment	33.3	excurrent
Inter-2°		Proximal Course	34.1.1	parallel to major secondaries
		Length	34.2.1	<50% of subjacent secondary
		Distal Course	34.3.1	reticulating
		Vein Frequency	34.4.1	<1 perintercoital
3°		Intercostal 3° Vein Fabric	35.1.1.3	mixed percurrent
		Angle of Percurrent Tertiaries	35.1.2.2	obtuse
		Vein Angle Variability	36.4	decreasing exmedially
		Epimedial Tertiaries	37.1.1.1	opposite percurrent
		Admedial Course	37.2.1.3	perpendicular to midvein
		Exmedial Course	37.2.2.1	parallel to intercostal tertiary
		Exterior Tertiary Course	38.3	terminating at the margin
4°		Quaternary Vein Fabric	39.2.2	irregular reticulate
5°		Quinternary Vein Fabric	40.1.2	irregular reticulate
		Areolation	41.2.2	moderate development
		FEV branching	42.1.1	FEVs absent
		FEV termination	99	n/a
		Marginal Ultimate Venation	43.1	absent

III. Teeth	Score	Description
Tooth Spacing	44.2	irregular
Number of Orders of Teeth	45.2	two
Teeth / cm		2
Sinus Shape	47.1	angular
Tooth Shapes		cc/cc
Tooth Shapes		
Tooth Shapes		
Tooth Shapes		
Tooth Shapes		
Principal Vein	49.1	present
Principal Vein Termination	51.2	marginal
Course of Accessory Vein	51.1	convex
Features of the Tooth Apex	53.1	none

Text Description:

Leaf attachment petiolate. Blade attachment marginal, laminar size microphyll to mesophyll, laminar L:W ratio 1.3:1, laminar shape ovate, blade medially asymmetrical, basal width asymmetrical, palmately lobed, margin serrate. Apex angle acute, apex shape acuminate to convex, base angle reflex, base shape cordate. Surface texture pubescent. Primary vein basal actinodromous, naked basal veins absent, five basal veins, simple agrophic veins, major 2° veins semicraspedodromous, interior secondaries absent, minor secondary course semicraspedodromous, major secondary spacing irregular, secondary angle smoothly decreasing proximally, major secondary attachment excurrent. Intersecondary length <50% of subjacent secondary, proximal course parallel to subjacent major secondary, distal course reticulating, vein frequency <1 per intercostal area. Intercostal tertiary vein fabric mixed percurrent with obtuse angle that decreases exmedially. Epimedial tertiaries opposite percurrent, proximal course perpendicular to midvein, distal course parallel to intercostal tertiary. Exterior tertiary course terminating at the margin. Quaternary and quinternary vein fabric irregular reticulate, areolation development moderate. Tooth spacing irregular, two orders of teeth, 2 teeth/cm, sinus shape rounded, tooth shape concave/concave. Principal vein present with termination at apex of tooth, accessory vein convex.

Example 16. Anacardiaceae - *Comocladia dodonaea*

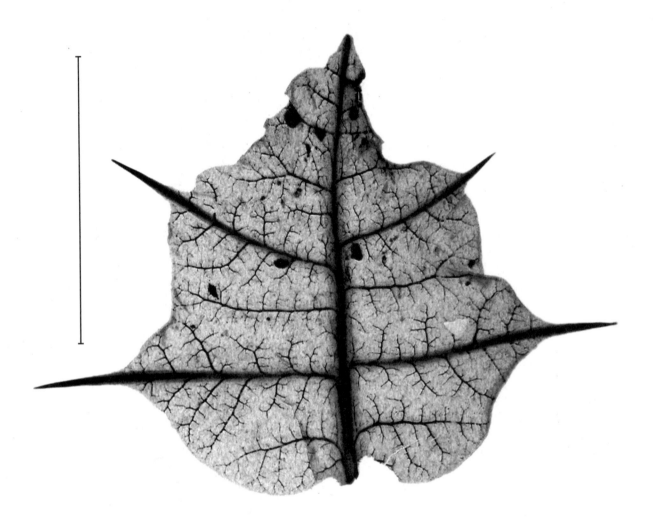

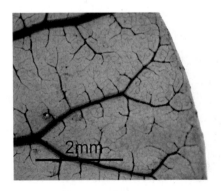

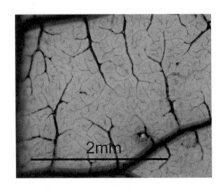

Anacardiaceae - *Comocladia dodonaea*

I. Leaf Characters	Score	Description
Leaf Attachment	1.1	petiolate
Leaf Arrangement	2.1	alternate
Leaf Organization	3.2.2.1	once pinnately compound
Leaflet Arrangement	4.3.1	opposite-odd
Leaflet Attachment	5.1	petiolulate
Petiole Features	88	not visible

Features of the Blade

	Score	Description
Position of Blade Attachment	7.1	marginal
Laminar Size	8.3	microphyll
Laminar L:W Ratio		1:1
Laminar Shape	10.3	ovate
Medial Symmetry	11.1	symmetrical
Base Symmetry	12.1	symmetrical
Base Symmetry	12.1	symmetrical
Lobation	13.1	unlobed
Margin Type	14.2	toothed
Special Margin Features	15.1.2	sinuous
Apex Angle	16.1	acute
Apex Shape	17.3	acuminate
Base Angle	18.3	reflex
Base Shape	19.2.1	cordate
Base Shape	19.2.1	cordate
Terminal Apex Features	88	not visible
Surface Texture	88	not visible
Surficial Glands	88	not visible

II. Venation		Score	Description
1°	Primary Vein Framework	23.1	pinnate
	Naked Basal Veins	24.1	absent
	Number of Basal Veins		3
	Agrophic Veins	0	absent
2°	Major 2° Vein Framework	27.1.1	craspedodromous
	Interior Secondaries	0	absent
	Minor Secondary Course	0	absent
	Perimarginal Veins	0	absent
	Major Secondary Spacing	31.1	regular
	Variation of Secondary Angle	32.3	smoothly increasing proximally
	Major Secondary Attachment	33.1	decurrent
Inter-2°	Proximal Course	34.1.1	parallel to major secondaries
	Length	34.2.2	>50% of subjacent secondary
	Distal Course	34.3.2	parallel to subjacent major secondary
	Vein Frequency	34.4.2	~1 per intercostal area
3°	Intercostal 3° Vein Fabric	35.3.4	transverse freely ramified
	Angle of Percurrent Tertiaries	99	n/a
	Vein Angle Variability	36.4	decreasing exmedially
	Epimedial Tertiaries	37.1.2	ramified
	Admedial Course	99	n/a
	Exmedial Course	99	n/a
	Exterior Tertiary Course	0	absent
4°	Quaternary Vein Fabric	0	absent
5°	Quinternary Vein Fabric	0	absent
	Areolation	41.2.1	poor development
	FEV branching	42.1.4.2	2 or more, dendritic
	FEV termination	42.2.2	tracheoid idioblasts
	Marginal Ultimate Venation	43.2	incomplete loops

III. Teeth	Score	Description
Tooth Spacing	44.1	regular
Number of Orders of Teeth	45.1	one
Teeth / cm		1
Sinus Shape	47.2	rounded
Tooth Shapes		
Tooth Shapes		
Tooth Shapes		
Tooth Shapes		
Principal Vein	49.1	present
Principal Vein Termination	51.2.1	at apex of tooth
Course of Accessory Vein	0	absent
Features of the Tooth Apex	53.3.1	spinose

Text Description:

Blade attachment marginal, laminar size microphyll, laminar L:W ratio 1:1, laminar shape ovate, blade medially symmetrical, base symmetrical, unlobed, margin toothed and sinuous. Apex angle acute, apex shape acuminate, base angle reflex, base shape cordate. Primary vein pinnate, naked basal veins absent, three basal veins, no agrophic veins. Major 2° veins craspedodromous, interior secondaries absent, minor secondaries absent, major secondary spacing regular, secondary angle smoothly increasing proximally, major secondary attachment decurrent. Intersecondary length >50% of subjacent secondary, proximal course parallel to major secondary, distal course parallel to subjacent major secondary, vein frequency ~1 per intercostal area. Intercostal tertiary vein fabric transverse freely ramified. Epimedial tertiaries ramified. Areolation development poor. Freely ending veinlets have two or more dendritic branches with tracheoid idioblasts. Marginal ultimate venation loops incompletely. Tooth spacing regular, one order of teeth, 1 tooth/cm, sinus shape rounded. Principal vein present with termination at apex of tooth, accessory vein absent. Tooth apex spinose.

Example 17. Anacardiaceae - *Sorindeia gilletii*

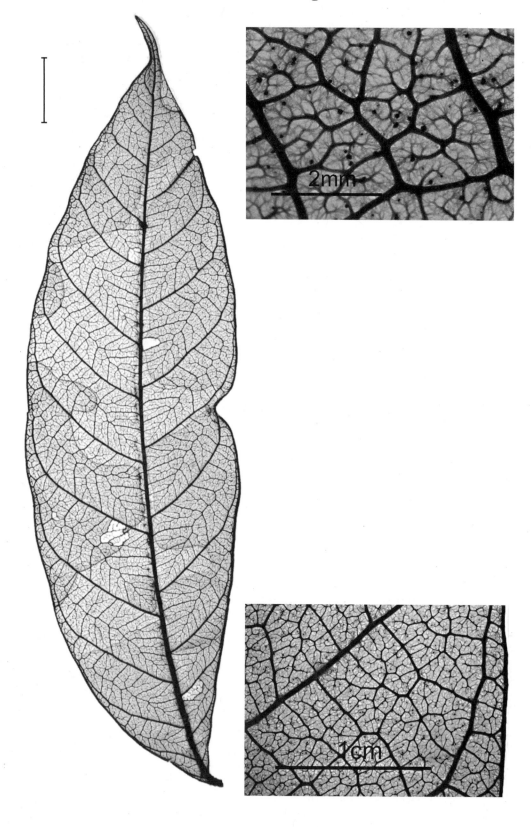

Anacardiaceae - *Sorindeia gilletii*

I. Leaf Characters	Score	Description
Leaf Attachment	1.1	petiolate
Leaf Arrangement	2.1	alternate
Leaf Organization	3.2.1.1	once pinnately compound
Leaflet Arrangement	4.1	alternate
Leaflet Attachment	5.1	petiolulate
Petiole Features	88	not visible

Features of the Blade

Position of Blade Attachment	7.1	marginal
Laminar Size	8.5	mesophyll
Laminar L:W Ratio		3.3:1
Laminar Shape	10.1	elliptic
Medial Symmetry	11.2	asymmetrical
Base Symmetry	12.2.1	basal width asymmetrical
Base Symmetry	12.2.3	basal insertion asymmetry
Lobation	12.2.3	unlobed
Margin Type	13.1	untoothed
Special Margin Features	0	absent
Apex Angle	16.1	acute
Apex Shape	17.3	acuminate
Base Angle	18.1	acute
Base Shape	19.1.1	straight
Base Shape	19.1.1	straight
Terminal Apex Features	0	absent
Surface Texture	88	not visible
Surficial Glands	88	not visible

	II. Venation	Score	Description
1°	Primary Vein Framework	23.1	pinnate
	Naked Basal Veins	24.1	absent
	Number of Basal Veins		1
	Agrophic Veins	0	absent
2°	Major 2° Vein Framework	27.2.1	eucamptodromous
	Interior Secondaries	28.1	absent
	Minor Secondary Course		n/a
	Perimarginal Veins	30.1	marginal secondary
	Major Secondary Spacing	31.1	regular
	Variation of Secondary Angle	32.1	uniform
	Major Secondary Attachment	33.3	excurrent
Inter-2°	Proximal Course	0	absent
	Length	99	n/a
	Distal Course	99	n/a
	Vein Frequency	99	n/a
3°	Intercostal 3° Vein Fabric	35.2.3	composite admedial
	Angle of Percurrent Tertiaries	99	n/a
	Vein Angle Variability	99	n/a
	Epimedial Tertiaries	37.1.3	reticulate
	Admedial Course		n/a
	Exmedial Course		n/a
	Exterior Tertiary Course	38.2	looped
4°	Quaternary Vein Fabric	39.2.2	irregular reticulate
5°	Quinternary Vein Fabric	40.2	freely ramifying
	Areolation	41.2.1	poor development
	FEV branching	42.1.4.2	2 or more, dendritic
	FEV termination	42.2.2	tracheoid idioblasts
	Marginal Ultimate Venation	43.1	absent

III. Teeth	Score	Description
Tooth Spacing	99	n/a
Number of Orders of Teeth	99	n/a
Teeth / cm	99	n/a
Sinus Shape	99	n/a
Tooth Shapes	99	n/a
Tooth Shapes	99	n/a
Tooth Shapes	99	n/a
Tooth Shapes	99	n/a
Principal Vein	99	n/a
Principal Vein Termination	99	n/a
Course of Accessory Vein	99	n/a
Features of the Tooth Apex	99	n/a

Text Description:

Blade attachment marginal, laminar size mesophyll, L:W ratio 3.3:1, laminar shape elliptic with medial asymmetry and basal width and insertion asymmetry. Margin is entire with acute apex angle, acuminate apex, acute base angle, and straight base shape. Primary venation is pinnate with no naked basal veins, one basal vein, and no agrophic veins. Major secondaries eucamptodromous with regular spacing, uniform angle, and excurrent attachment to midvein. Interior secondaries absent, minor secondaries absent, and marginal secondary present. Intersecondaries absent. Intercostal tertiary veins composite admedial. Epimedial tertiaries ramified. Exterior tertiaries reticulate. Quaternary vein fabric freely ramifying. Quinternary vein fabric freely ramifying. Areolation moderately developed. Freely ending veinlets have two or more dendritic branches with highly branched sclereids. Marginal ultimate venation absent.

Example 18. Proteales - *Leepierceia preartocarpoides*

Proteales fossil - *Leepierceia preartocarpoides*

I. Leaf Characters	Score	Description
Leaf Attachment	1.1	petiolate
Leaf Arrangement	88	not visible
Leaf Organization	88	not visible
Leaflet Arrangement	88	not visible
Leaflet Attachment	88	not visible
Petiole Features	88	not visible

Features of the Blade

	Score	Description
Position of Blade Attachment	7.1	marginal
Laminar Size	8.3	microphyll
Laminar L:W Ratio		1.75:1
Laminar Shape	10.1	elliptic
Medial Symmetry	11.1	symmetrical
Base Symmetry	12.1	symmetrical
Base Symmetry	12.1	symmetrical
Lobation	13.1	unlobed
Margin Type	14.2.2	serrate
Special Margin Features	0	absent
Apex Angle	16.1	acute
Apex Shape	17.3	acuminate
Base Angle	18.3	reflex
Base Shape	19.2.1	cordate
Base Shape	19.2.1	cordate
Terminal Apex Features	0	absent
Surface Texture	88	not visible
Surficial Glands	88	not visible

	II. Venation	Score	Description
1°	Primary Vein Framework	23.1	pinnate
	Naked Basal Veins	24.1	absent
	Number of Basal Veins		7
	Agrophic Veins	26.2.2	compound
2°	Major 2° Vein Framework	27.4	mixed
	Interior Secondaries	28.1	absent
	Minor Secondary Course	29.2	simple brochidodromous
	Perimarginal Veins	0	absent
	Major Secondary Spacing	31.1	regular
	Variation of Secondary Angle	32.1	uniform
	Major Secondary Attachment	33.3	excurrent
Inter-2°	Proximal Course	34.1.1	parallel to major secondaries
	Length	34.2.1	<50% of subjacent secondary
	Distal Course	34.3.2	parallel to subjacent major secondary
	Vein Frequency	34.4.1	<1 per intercostal area
3°	Intercostal 3° Vein Fabric	35.1.1.1	opposite percurrent
	Angle of Percurrent Tertiaries	35.1.2.2	obtuse
	Vein Angle Variability	36.3.1	basally concentric
	Epimedial Tertiaries	37.1.1.1	opposite percurrent
	Admedial Course	37.2.1.3	perpendicular to midvein
	Exmedial Course	37.2.2.1	parallel to intercostal tertiary
	Exterior Tertiary Course	38.2	looped
4°	Quaternary Vein Fabric	39.1.3	mixed percurrent
5°	Quinternary Vein Fabric	88	not visible
	Areolation	41.2.2	moderate development
	FEV branching	88	not visible
	FEV termination	88	not visible
	Marginal Ultimate Venation	88	not visible

III. Teeth	Score	Description
Tooth Spacing	44.2	irregular
Number of Orders of Teeth	45.1	one
Teeth / cm		0.2
Sinus Shape	47.2	rounded
Tooth Shapes		st/st
Tooth Shapes		cv/cv
Tooth Shapes		cc/cv
Tooth Shapes		
Principal Vein	49.1	present
Principal Vein Termination	51.2.1	at apex of tooth
Course of Accessory Vein	51.1.1	looped
Features of the Tooth Apex	53.1	none

Text Description:

Leaf attachment petiolate. Blade attachment marginal, laminar size microphyll, laminar L:W ratio 1.75:1, laminar shape elliptic, blade medially symmetrical, base symmetrical, unlobed, margin serrate. Apex angle acute, apex shape acuminate, base angle reflex, base shape cordate. Primary vein pinnate, naked basal veins absent, seven basal veins, compound agrophic veins. Major 2° veins simple brochidodromous, interior secondaries absent, minor secondaries brochidodromous, major secondary spacing regular, secondary angle uniform, major secondary attachment excurrent. Intersecondary length <50% of subjacent secondary, proximal course parallel to major secondary, distal course parallel to subjacent major secondary, vein frequency <1 per intercostal area. Intercostal tertiary vein fabric opposite percurrent with obtuse vein angle that is basally concentric. Epimedial tertiaries opposite percurrent with proximal course perpendicular to midvein and distal course parallel to intercostal tertiary. Exterior tertiary course looped. Quaternary vein fabric mixed percurrent, areolation development moderate. Tooth spacing irregular, one order of teeth, 0.2 teeth/ cm, sinus shape rounded, tooth shape straight/straight to concave/convex. Principal vein present with termination at apex of tooth.

Appendix C. Vouchers

Most of the images are from the National Cleared Leaf Collection, Department of Paleobiology, National Museum of Natural History, Smithsonian Institution. Slides prefixed with NCLC-W are housed at the Smithsonian; slides prefixed with NCLC-H are currently on long-term loan at the Yale Peabody Museum. Additional abbreviations include NYBG for the New York Botanical Garden, USNM for the Smithsonian (fossil) collection and DMNH for the Denver Museum of Nature & Science collection.

We used the Angiosperm Phylogeny Group (APG) website for family alignments (http://www.mobot.org/MOBOT/research/APweb/). When nomenclature was in doubt we used the International Plant Names Index (IPNI) (http://www.ipni.org/), and to a lesser extent, TROPICOS (http://mobot.mobot.org/W3T/Search/vast.html).

Fig.	Family	Genus and species	Collector and field number (where collected) slide no.
3	Dipterocarpaceae	*Dipterocarpus verrucosus* Foxw. ex v. Slooten	A. D. E. Elmer 21650 (Brunei) NCLC-W 1655
6	Iteaceae	*Itea chinensis* Hook. & Arn.	Peng 12615 (China) NCLC-H 3199
7	Berberidaceae	*Berberis sieboldii* Miq.	RWC (Japan) NCLC-W 450
9	Cornaceae	*Alangium chinense* (Lour.) Harms	K. King 1926 (Kiangsu, China) NCLC-W 1225
11	Altingiaceae	*Liquidambar styraciflua* L.	Ruth 264 (Tennessee, USA) NCLC-H 815
12	Euphorbiaceae	*Acalphya pringlei* S. Watson	T. H. Kearney 8000 (Pima Co., Arizona) NCLC-H 6185
15	Fabaceae	*Andira* sp.	D. Daly (Madre de Dios, Peru) unvouchered
16	Rosaceae	*Malus mandshurica* (Maxim.) Kom. ex Juz.	New York Botanical Garden living collection (USA)
18	Malvaceae	*Tilia chingiana* Hu & Cheng	Chung 434 (Kiangsi, China) NCLC-W 8629

Fig.	Family	Genus and species	Collector and field number (where collected) slide no.
19	Euphorbiaceae	*Discocleidion rufescens* (Fr.) Pax & K. Hoffm.	W. Y. Chun 5021 (Hupeh, China) NCLC-W 3022
20	Euphorbiaceae	*Croton lobatus* L.	L. Krapovickas & Cristóbal 12728 (Chaco, Argentina) NCLC-W 11584
21	Euphorbiaceae	*Aleurites remyi* Sherff	O. Degener 27455 (Hawaii, USA) NCLC-H 709
31	Fabaceae-Caesalpinioideae	*Hymenaea courbaril* L.	Cult. UCSC 259 (Puerto Rico) NCLC-W 4284
51	Fabaceae-Mimosoideae	*Acacia mangium* Willd.	D. W. Stevenson s.n. (Vinh Phuc Province, Vietnam)
52	Berberidaceae	*x Mahoberberis neubertii* C. K. Schneid.	s.n. (North Dakota, USA) NCLC-H 1175
53	Cabombaceae	*Brasenia schreberi* J. F. Gmel.	Deming s.n. (Connecticut, USA) NCLC-H 6693
54	Euphorbiaceae	*Macaranga bicolor* Müll. Arg.	E. D. Merrill 1533 (Luzon, Philippines) NCLC-W 854
56	Cucurbitaceae	*Trichosanthes formosana* Hayata	A. Henry 1952 (Taiwan) NCLC-H 2050
57	Menispermaceae	*Dioscoreophyllum strigosum* Engl.	J. Lebrun 2926 (Angodia, Congo) NCLC-W 7814
58	Celastraceae	*Cheiloclinium anomalum* Miers	B. A. Krukoff 6652 (Amazonas, Brazil) NCLC-W 8251
59	Apocynaceae	*Alstonia congensis* Engl.	W. T. S. Brown 2355 (Ghana) NCLC-W 5077
60	Chrysobalanaceae	*Parinari* sp.	(without collector) NCLC-W 12331
61	Moraceae	*Ficus citrifolia* Mill.	H. S. Irwin et al. (9/10/1960) (Amapá, Brazil) NCLC-W 10841
62	Proteaceae	*Xylomelum angustifolium* Kipp. ex Meissn.	C. C. Fauntleroy 2/17 (New South Wales, Australia) NCLC-W 6921
63	Celastraceae	*Maytenus aquifolium* Mart.	Y. Mexia 5241 (Minas Gerais, Brazil) NCLC-W 13582

Fig.	Family	Genus and species	Collector and field number (where collected) slide no.
64	Fabaceae-Faboideae	*Ramirezella pringlei* Rose	C. G. Pringle 13822 (Guerrero, Mexico) NCLC-W 14813
65	Euphorbiaceae	*Aleurites remyi* Sherff	O. Degener 27455 (Hawaii, USA) NCLC-H 709
66	Salicaceae	*Lunania mexicana* Brandeg.	J. A. Steyermark 47912 (Guatemala) NCLC-H 1838
67	Malvaceae	*Tilia chingiana* Hu & Cheng	Chung 434 (Kiangsi, China) NCLC-W 8629
68	Oleaceae	*Fraxinus floribunda* Wallich	E. H. Wilson 2786 (Hupeh, China) NCLC-W 8963
69	Chrysobalanaceae	*Parinari campestris* Aubl.	G. T. Prance s.n. (Brazil) NCLC-H 4003
70	Euphorbiaceae	*Melanolepis multiglandulosa* Rchb. & Zoll.	E. D. Merrill 489 (Blanco, Philippines) NCLC-W 871
71	Passifloraceae	*Adenia heterophylla* (Blume) Koord.	E. D. Merrill 5958 (Philippines) NCLC-H 1935
72	Rosaceae	*Potentilla recta* Jacq.	Davidson 4197 (Iowa, USA) NCLC-H 3897
73	Proteaceae	*Stenocarpus sinuatus* Endl.	J. Wolfe, 1974 (cult. Missouri, USA, U815) NCLC-W 10238
74	Proteaceae	*Dryandra longifolia* R. Br.	Harvey Herb. (cult. Paris) NCLC-W 6334
75	Cucurbitaceae	*Cucurbita cylindrata* L. H. Bailey	I. L. Wiggins s.n. (Mexico) NCLC-H 2051
76	Fabaceae-Caesalpinioideae	*Bauhinia madagascariensis* Desv.	Brion (1843) (Madagascar) NCLC-W 5733
77	Clusiaceae	*Caraipa punctulata* Ducke	A. Ducke 35410, (Brazil) NCLC-H 1832
78	Salicaceae	*Casearia ilicifolia* Vent.	Miller 276 (Haiti) NCLC-H 1061
79	Betulaceae	*Betula lenta* L.	J. U. McClammer s.n. (Virginia, USA) NCLC-H 5415

Fig.	Family	Genus and species	Collector and field number (where collected) slide no.
80	Violaceae	*Viola brevistipulata* W. Becker	H. Koidzumi 836 (Japan) NCLC-H 2108
81	Fagaceae	*Quercus alba* L.	E. Kowalski and D. Dilcher 126/132 (Millbrook, NY)
82	Rosaceae	*Rubus mesogaeus* Focke ex Diels	J. F. C. Rock 8636 (Yunnan, China) NCLC-W 12100
83	Phyllanthaceae	*Bridelia cathartica* Bertol.f.	C. E. Tanner 3541 (Tanzania) NCLC-W 11529
88	Betulaceae	*Ostrya guatemalensis* Rose	Le Sueur 1305 (Chihuahua, Mexico) NCLC-W 6773
89	Berberidaceae	*x Mahoberberis neubertii* C. K. Schneid.	(without collector) (North Dakota, USA) NCLC-H 1175
90	Fabaceae-Caesalpinioideae	*Bauhinia madagascariensis* Desv.	Brion (1843) (Madagascar) NCLC-W 5733
91	Menispermaceae	*Dioscoreophyllum strigosum* Engl.	J. Lebrun 2926 (Angodia, Congo) NCLC-W 7814
93	Elaeocarpaceae	*Aristotelia racemosa* Hook. f.	Anderson 260 (South Island, New Zealand) NCLC-W 9487
94	Actinidiaceae	*Saurauia calyptrata* Lauterb.	L. J. Brass 10908 (Papua New Guinea) NCLC-W 8944
95	Anacardiaceae	*Ozoroa obovata* (Oliv.) R. Fern. & A. Fern.	A. Moura 43 (Mozambique) NCLC-W 10067
96	Magnoliaceae	*Liriodendron chinense* (Hemsl.) Sarg.	Chaney s.n. (Kiangsu, China) NCLC-W 1553a
97	Annonaceae	*Neouvaria acuminatissima* (Miq.) Airy-Shaw	A. D. E. Elmer 21112 (Tawao, Philippines) NCLC-W 7851
98	Hamamelidaceae	*Corylopsis veitchiana* Bean	JAW (7/6/64) (cult. Royal Botanic Gardens, Kew) NCLC-W 1126
99	Bignoniaceae	*Lundia spruceana* Bur.	J. Steinbach 7333 (Santa Cruz, Bolivia) NCLC-W 218
100	Dichapetalaceae	*Tapura guianensis* Aubl.	Wachenheim (6/23/21) (French Guiana) NCLC-W 8070

Fig.	Family	Genus and species	Collector and field number (where collected) slide no.
101	Dilleniaceae	*Schumacheria castaneifolia* Vahl	S. Sohmer s.n. (Sri Lanka) NCLC-H 6793
102	Anacardiaceae	*Mauria heterophylla* Kunth	Rimbach 38 (Ecuador) NCLC-W4218
103	Malvaceae	*Tilia chingiana* Hu & Cheng	Chung 434 (Kiangsi, China) NCLC-W 8629
104	Aristolochiaceae	*Asarum europaeum* L.	Hawes s.n. (Poland) NCLC-H 6692
105	Menispermaceae	*Cissampelos owariensis* Beauv. ex DC. (= *C. pareira* L.)	Gilbert 2045 (Congo) NCLC-W 4498
106	Juglandaceae	*Carya leiodermis* Sarg.	W. Wolf (Alabama, USA) NCLC-W 8484
107	Lauraceae	*Sassafras albidum* (Nutt.) Nees	W. B. Marshall s.n. (New Jersey, USA) NCLC-W 6281
108	Rosaceae	*Prunus mandshurica* (Maxim.) Koehne	E. H. Wilson 8775 (Korea) NCLC-W 8775
109	Apocynaceae	*Carissa opaca* Stapf. ex Haines	U. Singh 136 (India) NCLC-W 13732
110	Salicaceae	*Populus dimorpha* Brandeg.	M. E. Jones (2-3-27) (Sinaloa, Mexico) NCLC-W 1262
111	Menispermaceae	*Diploclisia chinensis* Merr.	Metcalf 2296 (Fukier) NCLC-W 242
112	Euphorbiaceae	*Adelia triloba* Hemsl.	Steyermark 17489 (Panama) NCLC-W 2928
113	Apocynaceae	*Alstonia plumosa* Labill.	O. Degener 14673 (Fiji) NCLC-W 13703
114	Berberidaceae	*Berberis sieboldii* Miq.	R. W. Chaney (Japan) NCLC-W 450
115	Phyllanthaceae	*Phyllanthus poumensis* Guillaumin	H. McKee 4620 (New Caledonia) NCLC-W 11758
116	Cercidiphyllaceae	*Cercidiphyllum japonicum* Sieb. & Zucc.	(without collector) NCLC-W 9085

Fig.	Family	Genus and species	Collector and field number (where collected) slide no.
117	Sapindaceae	*Acer saccharinum* L.	Knowlton s.n. (Maine, USA) NCLC-H 6861
118	Hamamelidaceae	*Liquidambar styraciflua* L.	H. Meyer (cult. Strybing Arb. 66-125) NCLC-W 11912
119	Alismataceae	*Sagittaria* sp.	(without collector) NCLC-W 797
120	Apocynaceae	*Araujia angustifolia* Steud.	Palacios-Cuezzo 2233 (Corrientes, Argentina) NCLC-W 10244
121	Menispermaceae	*Cocculus ferrandianus* Gaudich.	Kruckeberg 97 (Oahu, Hawaii, USA) NCLC-W 10432
122	Fabaceae	*Bauhinia rubeleruziana* J. D. Smith	(without collector) NCLC-W 30221
123	Annonaceae	*Fitzalania heteropetala* F. Muell.	F. von Mueller (Port Dennison, Australia) NCLC-W 14543
131	Anacardiaceae	*Astronium graveolens* Jacq.	B. Wallnöfer 9567 (Peru) NY
133	Betulaceae	*Carpinus fargesii* C. K. Schneid.	Li 13081 (Anhui, China) NCLC-H 6455
135	Bixaceae	*Bixa orellana* L.	J. Cuatrecasas 7403 (Colombia) NCLC-H 6255
136	Anacardiaceae	*Comocladia cuneata* Britton (syn.: *C. acuminata*)	R. A. & E. S. Howard 8249 (Dominican Republic) NCLC-W 8197
138	Griseliniaceae	*Griselinia scandens* Taub.	E. Werdermann 923 (Coquimbo, Chile) NCLC-W 6513
139	Ranunculaceae	*Delphinium cashmerianum* Royle	(without collector) (Calcutta, India) NCLC-H 1477
140	Sapindaceae (ex-Aceraceae)	*Acer argutum* Maxim.	E. H. Wilson 1914 (Botanic Garden Sapporo, Japan) NCLC-W 8578
141	Fagaceae	*Fagus longipetiolata* Seemen	C. T. Hwa 36 (Szechuan, China) NCLC-W 1412
142	Melastomataceae	*Meriania speciosa* (Bonpl.) Naudin	N. Espinal 3533 (Colombia) NCLC-W 9286

Fig.	Family	Genus and species	Collector and field number (where collected) slide no.
143	Melastomataceae	*Loreya arborescens* (Aubl.) DC. (syn.: *L. acutifolia*)	R. E. Schultes & Cabrera 19755 (Amazonas, Colombia) NCLC-W 9280
144	Euphorbiaceae	*Givotia rottleriformis* Griff. ex Wight	R. G. Cooray 69100203R (Sri Lanka) NCLC-W 9046
145	Euphorbiaceae	*Tannodia swynnertonii* Prain	J. McGregor M4/48 (Zimbabwe) NCLC-W 4631
146	Trochodendraceae	*Tetracentron sinense* Oliv.	(without collector) (China) NCLC-H 184
147	Clusiaceae	*Calophyllum calaba* L.	Hahm 150 (Martinique) NCLC-W 4372
148	Lauraceae	*Sassafras albidum* (Nutt.) Nees	R. K. Godfrey & Tryon 1443 (S Carolina, USA) NCLC-H 6280
149	Hamamelidaceae	*Parrotia jacquemontiana* Decne.	JAW (7/6/64) (cult. Royal Botanic Gardens, Kew) NCLC-W 1128
151	Buxaceae	*Buxus glomerata* (Griseb.) Müll. Arg.	A. H. Liogier 11086 (Dominican Republic) NCLC-H 6247
152	Euphorbiaceae	*Croton hircinus* Vent.	H. H. Pittier 5025 (Panama) NCLC-H 6223
153	Anacardiaceae	*Spondias globosa* J. D. Mitch. & Daly	D. C. Daly et al. 7836 (Acre, Brazil) NY
154	Menispermaceae	*Diploclisia kunstleri* (King) Diels	(without collector) Nat. Col., B. Sci. 2175 (Sarawak) NCLC-W 8815
155	Betulaceae	*Ostrya guatemalensis* Rose	P. Grant (May 22, 1963) (Nayarit, Mexico) NCLC-W 14869
156	Salicaceae	*Carrierea calycina* Franch.	E. H. Wilson 3227 (W. China [4000 ft.]) NCLC-W 7957
157	Euphorbiaceae	*Dalechampia cissifolia* Poepp. & Endl.	D. M. Porter et al. 4887 (Panama) NCLC-W 11597
158	Euphorbiaceae	*Croton hircinus* Vent.	H. H. Pittier 5025 (Panama) NCLC-H 6223
159	Malvaceae	*Dombeya elegans* Cordem.	J. Wolfe (6/26/64) (cult. Royal Botanic Gardens, Kew) NCLC-W 1170

Fig.	Family	Genus and species	Collector and field number (where collected) slide no.
160	Datiscaceae	*Tetrameles nudiflora* R. Br.	M. Jayasuriya 1336 (Sri Lanka) NCLC-H 4947
161	Lauraceae	*Phoebe costaricana* Mez & Pittier	M. E. Derdson 583 (Panama) NCLC-W 5550
162	Platanaceae	*Platanus racemosa* Nutt.	J. Wolfe (Berkeley, Calif., USA) NCLC-W 500
163	Cucurbitaceae	*Trichosanthes formosana* Hayata	A. Henry 1952 (Taiwan) NCLC-H 2050
164	Rhamnaceae	*Paliurus ramosissimus* Poir.	Tsang 27855 (Kwangsin, China) NCLC-W 1796
165	Piperaceae	*Sarcorhachis naranjoana* Trel.	R. Lent 1586 (Costa Rica) NCLC-W 12667
166	Melastomataceae	*Topobea watsonii* Cogn.	E. Contreras 6168 (Guatemala) NCLC-W 7585
167	Proteaceae	*Paranomus sceptrum* Kuntze	N. S. Pillans 10899 (South Africa) NCLC-W 5246
168	Potamogetonaceae	*Potamogeton amplifolius* Tuckerm.	Pick s.n. (Oregon, USA) NCLC-H 6777
169	Ruscaceae	*Maianthemum dilatatum* (Wood.) A. Nelson & J. F. Macbr.	L. Roush (7/6/1919) (Washington State, USA) NCLC-W 17896
170	Cucurbitaceae	*Trichosanthes formosana* Hayata	A. Henry 1952 (Taiwan) NCLC-H 2050
171	Sapindaceae	*Acer miyabei* Maxim	(without collector) NCLC-W 9072
172	Rosaceae	*Sorbus japonica* (Decne.) Hedlund	Shiota 6315 (Mino, Japan) NCLC-W 8671
173	Cunoniaceae	*Eucryphia glandulosa* Reiche	Aravena (Linares, Chile) NCLC -W 2468
174	Euphorbiaceae	*Alchornea tiliifolia* Müll. Arg.	A. Henry 12131C (China) NCLC-H 406
175	Hamamelidaceae	*Parrotia jacquemontiana* Decne.	J. Wolfe (7/6/1964) (cult. Royal Botanic Gardens, Kew) NCLC-W 1128

Fig.	Family	Genus and species	Collector and field number (where collected) slide no.
176	Vitaceae	*Cissus caesia* Afzel.	Melville & Hooker 461 (Sierra Leone) NCLC-W 4948
177	Hamamelidaceae	*Corylopsis glabrescens* Franch. & Sav.	Walker 7663 (Pennsylvania, USA) NCLC-H 821
178	Desfontaineaceae	*Desfontainea spinosa* Ruiz & Pav.	J. Cuatrecasas 11814 (Colombia) NCLC-H 4085
179	Menispermaceae	*Cyclea merrillii* Diels	J. Clemens 16713 (Luzon, Philippines) NCLC-W 4036
180	Salicaceae	*Aphaerema spicata* Miers	P. K. H. Dusén (11/25/14) (Paraná, Brazil) NCLC-W 1570
181	Cercidiphyllaceae	*Cercidiphyllum japonicum* Sieb. & Zucc.	R. W. Chaney s.n. (Japan) NCLC-W 26
182	Salicaceae	*Casearia ilicifolia* Vent.	Miller 276 (Haiti) NCLC-H 1061
183	Atherospermataceae	*Laurelia novae-zelandiae* A. Cunn.	A. K. Meebold s.n. (New Zealand) NCLC-H 6724
184	Trochodendraceae	*Tetracentron sinense* Oliv.	W. P. Fang 6705 (Szechuan, China) NCLC-W 6550
185	Berberidaceae	*Mahonia wilcoxii* (Kearney) Rehder	R. S. Ferris 9991 (Arizona, USA) NCLC-W 15043
186	Dilleniaceae	*Tetracera rotundifolia* Sm.	Idrobo & Schultes 1320 (Colombia) NCLC-H 831
187	Cornaceae	*Cornus officinalis* Sieb. & Zucc.	Li 13101 (Anhui, China) NCLC-H 6496
188	Dipterocarpaceae	*Isoptera lissophylla* Liv.	de Silva 53 (Sri Lanka) NCLC-W 1662
189	Melastomataceae	*Tococa aristata* Benth.	H. A. Gleason 625 (Guyana) NCLC-W 9296
190	Phyllanthaceae	*Cleistanthus oligophlebius* Merr.	A. D. E. Elmer 21651 (Brunei) NCLC-W 11559
191	Rhamnaceae	*Rhamnidium elaeocarpum* Reiss.	Pereira s.n. (Brazil) NCLC-H 4811

Fig.	Family	Genus and species	Collector and field number (where collected) slide no.
192	Cunoniaceae	*Eucryphia moorei* F. Muell.	R. Schodde 3496 (New South Wales, Australia) NCLC-W 2470
193	Anacardiaceae	*Cotinus obovatus* Raf.	W. Hess et al. 7511 (USA) NY
194	Phyllanthaceae	*Baccaurea staudtii* Pax	G. A. Zenker 568 (Cameroon) NCLC-H 11493
195	Burseraceae	*Santiria samarensis* Merr.	(without collector) (Camaris, Philippines) NCLC-H 208
196	Aextoxicaceae	*Aextoxicon punctatum* Ruiz & Pav.	(without collector) B. Sparre & Constance 10742 (Osorno, Chile) NCLC-W 2932
197	Polygonaceae	*Antigonon cinerascens* M. Martens & Galeotti	A. Ventura 4342 (Veracruz, Mexico) NCLC-W 14958
198	Canellaceae	*Capsicodendron pimenteira* Hoehne	P. K. H. Dusén 1033a (Paraná, Brazil) NCLC-H 238
199	Dichapetalaceae	*Tapura guianensis* Aubl.	Wachenheim s.n., 6/23/21 (French Guiana) NCLC-W 8070
200	Anacardiaceae	*Comocladia glabra* (Schult.) Spreng.	A. H. Liogier et al. 32748 (Puerto Rico) NY
201	Rosaceae	*Filipendula occidentalis* Howell	L. F. Henderson (7/11/1882) (Oregon, USA) NCLC-W 10707
202	Malvaceae	*Triplochiton scleroxylon* K. Schum.	A. J. M. Leeuwenberg 2877 (Ivory Coast) NCLC-W 3656
203	Achariaceae	*Scaphocalyx spathacea* Ridl.	(without collector) (Malaysia) NCLC-H 953
204	Adoxaceae	*Viburnum setigerum* Hance	Li 13015 (Anhui, China) NCLC-H 6461
205	Bixaceae	*Bixa orellana* L.	J. Cuatrecasas 7403 (Colombia) NCLC-H 6255
206	Asteraceae	*Philactis zinnioides* Schrad.	King 3446 (Chiapas, Mexico) NCLC-W 15130
207	Polygalaceae	*Securidaca marginata* Benth.	G. T. Prance et al. s.n. (Brazil) NCLC-H 2679

Fig.	Family	Genus and species	Collector and field number (where collected) slide no.
208	Anacardiaceae	*Spondias bivenomarginalis* K. M. Feng & P. Y. Mao	Liu Xingqi 27277 (China) MO
209	Melastomataceae	*Graffenrieda anomala* Triana	J. Cuatrecasas 16820 (Valle, Colombia) NCLC-W 9273
210	Fagaceae	*Castanea sativa* Mill.	(without collector) (Washington, DC, USA) NCLC-H 1441
211	Lamiaceae	*Vitex limonifolia* Wall.	R. M. King 5488 (Kanchanaburi, Thailand NCLC-W 6656
212	Proteaceae	*Kermadecia sinuata* Brongn. & Gris	M. Mackee 12877 (New Caledonia) NCLC-W 6599
213	Phyllanthaceae	*Glochidion bracteatum* Gillespie	A. C. Smith 7366 (Fiji) NCLC-W 11666
214	Salicaceae	*Populus jackii* Sarg.	Mairie-Victorin (6/30/33) (Montreal, Canada) NCLC-W 1265
215	Malvaceae	*Apeiba macropetala*	A. Ducke 18080 (Rio de Janeiro, Brazil) NCLC-H 5343
216	Malvaceae	*Tilia heterophylla* Vent.	J. Bright 9369 (Pennsylvania, USA) NCLC-W 7734
217	Euphorbiaceae	*Alchornea polyantha* Pax & K. Hoffm.	F. C. Lehmann (Cauca, Colombia) USNH 1856534
218	Moraceae	*Pseudolmedia laevis* Ruiz & Pav.	B. A. Krukoff 10256 (La Paz, Bolivia) NCLC-W 10906
219	Annonaceae	*Popowia congensis* Engl. & Diels	Louis 724 (Congo) NCLC-W 5442
220	Malpighiaceae	*Banisteriopsis laevifolia* (A. Juss.) B. Gates	Y. Mexia 5666 (Minas Gerais Brazil) NCLC-W 6553
221	Malvaceae	*Microcos tomentosa* Sm.	R. S. Toroes 1913 (Sumatra) NCLC-W 11503
222	Iteaceae	*Itea chinensis* Hook. & Arn.	Peng 12615 (China) NCLC-H 3199
223	Rosaceae	*Crataegus brainerdii* Sarg.	W. L. C. Muenscher & Lindsey 3373 (New York, USA) NCLC-W 11964

Fig.	Family	Genus and species	Collector and field number (where collected) slide no.
224	Dilleniaceae	*Tetracera podotricha* Gilg.	H. J. R. Vanderyst 25190 (Sanga, Congo) NCLC-W 7841
225	Cannabaceae	*Celtis cerasifera* C. K. Schneid.	G. Forrest 24471 (E Tibet/SW China) NCLC-W 9000
226	Chrysobalanaceae	*Couepia paraensis* (Mart. & Zucc.) Benth.	J. J. Wurdack & Monachino 39893 (Bolívar, Venezuela) NCLC-W 4142
227	Burseraceae	*Protium subserratum* (Engl.) Engl.	J. J. Pipoly & Gharbarran 10170 (Guyana) NY
228	Burseraceae	*Dacryodes negrensis* Daly & M. C. Martínez	G. T. Prance et al. 16147 (Amazonas, Brazil) NY
229	Burseraceae	*Protium opacum* Swart	B. A. Krukoff 4816 (Amazonas, Brazil) NCLC-W 13245
230	Burseraceae	*Santiria griffithii* Engl.	Anta 56 (Indonesia) NY
231	Anacardiaceae	*Comocladia cuneata* Britton (syn.: *C. acuminata* Britton)	R. A. & E. S. Howard 8249 (Dominican Republic) NCLC-W 8197 NY
232	Ancistrocladaceae	*Ancistrocladus tectorius* Merrill	H. Fung 20372 (Wen-Ch'ang, China) NCLC-H 5747
233	Burseraceae	*Canarium ovatum* Engl.	Molina 24514 (Philippines) NY
234	Dipterocarpaceae	*Stemonoporus nitidus* Thw.	P. S. Ashton 2003 (Sri Lanka) NCLC-H 4665
235	Meliaceae	*Guarea tuberculata* Vell.	D. Vincent (Brazil) NCLC-W 15406
236	Meliaceae	*Cedrela angustifolia* Moc. & Sessé ex DC.	Cooper & Slater (Panama) NCLC-H 640
237	Ochnaceae	*Ouratea* aff. *O. garcinioides* Ule	T. Lasser 58 (Venezuela) NCLC-H 5701
238	Violaceae	*Melicytus fasciger* Gillespie	H. U. Stauffer 5827 (Fiji) NCLC-W 3246
239	Dipterocarpaceae	*Note*: genus and species unknown	P. S. Ashton (s.n.) NCLC-H 4552

Fig.	Family	Genus and species	Collector and field number (where collected) slide no.
240	Acanthaceae	*Aphelandra pulcherrima* Kunth	J. Cuatrecasas & Castañeda 25012 (Magdalena, Colombia) NCLC-H 1207
241	Elaeocarpaceae	*Vallea stipularis* L.f.	E. L. Little 6129 (Pichincha, Ecuador) NCLC-H 5479
242	Burseraceae	*Santiria samarensis* Merr.	G. H. J. Wood 1791 (Brunei) NCLC-W 1733
243	Dilleniaceae	*Davilla rugosa* Poir.	E. G. Holt & Gehringer 413 (Amazonas, Venezuela) NCLC-H 845
245	Lauraceae	*Nectandra cuspidata* Nees & Mart. ex Nees	T. G. Tutin 465 (Papua New Guinea) NCLC-H 731
246	Elaeocarpaceae	*Sloanea eichleri* K. Schum.	L. O. Williams 13211 (Venezuela) NCLC-H 5369
247	Celastraceae	*Bhesa archboldiana* (Merr. & L. M. Perry) Ding Hou	L. J. Brass 28105 (Papua New Guinea) NCLC-H 4421
248	Picrodendraceae	*Piranhea trifoliata* Baill.	B. A. Krukoff 5924 (Pará, Brazil) NY, NCLC-W 4626
249	Huaceae	*Afrostyrax kamerunensis* Perkins & Gilg.	G. A. Zenker 365 (Cameroon) NCLC-W 3257
250	Anacardiaceae	*Sorindeia gilletii* De Wild.	J. M. & B. Reitsma 1420 (Gabon) NYBG
251	Anacardiaceae	*Protorhus nitida* Engl.	L. J. Dorr et al. 4617 (Madagascar) NYBG
252	Ochnaceae	*Ouratea thyrsoidea* Engl.	L. O. Williams 15365 (Venezuela) NCLC-H 5721
253	Anacardiaceae	*Comocladia glabra* (Schult.) Spreng.	A. H. Liogier et al. 32748 (Puerto Rico) NYBG
254	Anacardiaceae	*Rhus (Melanococca) taitensis* Guill.	T. G. Yuncker 9332 (Tahiti) NYBG
255	Adoxaceae	*Viburnum sempervirens* K. Koch	S. K. Lau 3991 (Kiangsi, China) NCLC-H 1365
256	Ebenaceae	*Diospyros maritima* Blume	E. D. Merrill 9340 (Philippines) NCLC-W 13192

Fig.	Family	Genus and species	Collector and field number (where collected) slide no.
257	Malvaceae	*Eriolaena malvacea* (H. Lév.) Hand.-Mazz	A. Henry 12506 B (Yunnan, China) NCLC-W 8045
258	Euphorbiaceae	*Macaranga bicolor* Müll. Arg.	E. D. Merrill 1533 (Luzon, Philippines) NCLC-W 854
259	Juglandaceae	*Juglans boliviana* Dode	Knowles & Bent s.n. (Metraro, Peru) NCLC-W 956b
260	Apocynaceae	*Odontadenia geminata* Müll. Arg.	Y. Mexia 6023 (Pará, Brazil) NCLC-W 9178
261	Salicaceae	*Flacourtia rukam* Zoll. & Mor.	Taj 638 (Hainan, China) NCLC-W 1577b
262	Actinidiaceae	*Actinidia latifolia* Merr.	A. Petelot 8649 (Vietnam) NCLC-W 8942
263	Cornaceae	*Alangium chinense* (Lour.) Harms	K. Ling (8/5/1926) (Kiangsu, China) NCLC-W 1225
264	Bixaceae	*Bixa orellana* L.	C. F. Baker 2000 (Nicaragua) NCLC-W 3234
265	Ochnaceae	*Ouratea thyrsoidea* Engl.	L. O. Williams 15365 (Venezuela) NCLC-H 5721
266	Berberidaceae	*Mahonia wilcoxii* Rehder	R. S. Ferris 9991 (Arizona, USA) NCLC-W 15043
267	Acanthaceae	*Aphelandra pulcherrima* Kunth	J. Cuatrecasas s.n. (Colombia) NCLC-H 1297
268	Capparaceae	*Capparis lundellii* Standl.	D. E. Breedlove 42274 (Chiapas, Mexico) NCLC-W 15061b
269	Dilleniaceae	*Dillenia indica* Blanco	R. Jaramillo & Dugand 4062 (Colombia) NCLC-H 918
270	Salicaceae	*Lunania mexicana* Brandeg.	C. A. Purpus 7381 (Chiapas, Mexico) NCLC-W 2693
271	Celastraceae	*Celastrus racemosus* Hayata	C. G. Pringle (Mexico) NCLC-H 4387
272	Elaeocarpaceae	*Sloanea eichleri* K. Schum.	L. O. Williams 13211 (Venezuela) NCLC-H 5369

Fig.	Family	Genus and species	Collector and field number (where collected) slide no.
273	Bixaceae	*Bixa orellana* L.	C. F. Baker (Nicaragua) NCLC-W 3234
274	Malvaceae	*Theobroma microcarpa* Mart.	B. A. Krukoff 1644 (Mato Grosso, Brazil) NCLC-W 3654
275	Connaraceae	*Spiropetalum erythrosepalum* Gilg. ex Schellen.	G. A. Zenker 584 (Cameroon) NCLC-W 4198
276	Burseraceae	*Commiphora aprevalii* Guillaumin	P. Phillipson 1814 (Madagascar) NYBG
277	Chloranthaceae	*Hedyosmum costaricense* C. E. Wood ex Burger	J. Luteyn & Stone 696 (Alajuela, Costa Rica) NCLC-H 6347B
278	Picramniaceae	*Picramnia krukovii* A. C. Sm.	B. A. Krukoff 5679 (Acre, Brazil) NCLC-W 13207
279	Monimiaceae	*Mollinedia floribunda* Tul.	Y. Mexia 5098 (Minas Gerais, Brazil) NCLC-W 10597
280	Lecythidaceae	*Barringtonia reticulata* Miq.	Escritor 21512 (Mindanao, Philippines) NCLC-W 12636
281	Apocynaceae	*Carissa bispinosa* Desf.	L. J. Brass 16182 (Malawi) NCLC-W 5044
282	Celastraceae	*Gymnosporia senegalensis* Loes.	Imperial Forest Institute 456 (Tanzania) NCLC-H 4441
283	Dipterocarpaceae	*Shorea congestiflora* (Thw.) P. S. Ashton	P. S. Ashton 2022 (Sri Lanka) NCLC-H 4636
284	Malvaceae	*Theobroma microcarpa* Mart.	B. A. Krukoff 6203 (Amazonas, Brazil) NCLC-H 5641
285	Cornaceae	*Alangium chinense* (Lour.) Harms	K. Ling (8/5/1926) (Kiangsu, China) NCLC-W 1225a
286	Huaceae	*Afrostyrax kamerunensis* Perkins & Gilg.	G. A. Zenker 365 (Cameroon) NCLC-W 3257
287	Ebenaceae	*Diospyros pellucida* Hiern	A. D. Elmer s.n. (Luzon, Philippines) NCLC-H 5100
288	Anacardiaceae	*Comocladia cuneata* Britton (syn.: *C. acuminata* Britton)	R. A. & E. S. Howard 8249 (Dominican Republic) NCLC-W 8197

Fig.	Family	Genus and species	Collector and field number (where collected) slide no.
289	Moraceae	*Pseudolmedia laevis* (Ruiz & Pav.) J. F. Macbr.	B. A. Krukoff 10256 (La Paz, Bolivia) NCLC-W 10906
290	Ebenaceae	*Diospyros hispida* A. DC.	H. S. Irwin 32089 (Brazil) NCLC-H 5022
291	Dipterocarpaceae	*Stemonoporus nitidus* Thw.	P. S. Ashton 2003 (Sri Lanka) NCLC-H 4665
292	Anacardiaceae	*Rhus (Melanococca) taitensis* Guill.	T. G. Yuncker 9332 (Tahiti) NYBG
293	Chloranthaceae	*Chloranthus glaber* (Thunb.) Makino	Tsang 21487 (Kwangtung, China) NCLC-W 2329
294	Clusiaceae	*Clusiella pendula* Cuatrec.	Killip 34966 (Colombia) UCH967992 NCLC-W 2648
295	Picrodendraceae	*Piranhea trifoliata* Baill.	B. A. Krukoff 5924 (Pará, Brazil) NY, NCLC-W 4626
296	Huaceae	*Afrostyrax kamerunensis* Perkins & Gilg.	G. A. Zenker 365 (Cameroon) NCLC-W 3257
302	Violaceae	*Melicytus fasciger* Gillespie	Stauffer 5827 (Fiji) NCLC-W 3246
303	Burseraceae	*Bursera inaguensis* Britton	G. V. Nash & N. Taylor 1205 (Bahamas) NYBG
304	Burseraceae	*Tetragastris panamensis* (Engl.) Kuntze	S. A. Mori et al. 14969 (French Guiana) NYBG
305	Euphorbiaceae	*Pycnocoma littoralis* Pax	A. V. Bogdan VB 622 (Kenya) NCLC-W 3141
308	Monimiaceae	*Mollinedia floribunda* Tul.	Y. Mexia 5098 (Minas Gerais, Brazil) NCLC-W 10597
309	Picramniaceae	*Picramnia krukovii* A. C. Sm.	B. A. Krukoff 5679 (Acre, Brazil) NCLC-W 13207
310	Phyllanthaceae	*Aporusa frutescens* Blume	Ramos 1364 (Brunei) NCLC-W 11487
312	Rhamnaceae	*Gouania velutina* Reiss.	H. Rombouts 662 (Brazil) NCLC-H 5324

Fig.	Family	Genus and species	Collector and field number (where collected) slide no.
313	Hydrangeaceae	*Dichroa philippinensis* Schltr.	A. D. E. Elmer 16177 (Luzon, Philippines) NCLC-W 2161
314	Celastraceae	*Campylostemon mucronatum* (Exell) J. B. Hall	A. J. M. Leeuwenberg 4118 (Ivory Coast) NCLC-W 6867
315	Vitaceae	*Leea macropus* Lauterb. & K. Schum.	JAW (6/26/64) (cult. Royal Botanic Gardens, Kew) NCLC-W 1151
316	Elaeocarpaceae	*Aristotelia racemosa* Hook.f.	L. Hickey s.n. (New Zealand) NCLC-H 6479
317	Rosaceae	*Crataegus brainerdi* Sarg.	Muenscher & Lindsey 3373 (New York, USA) NCLC-W 11964
318	Hydrangeaceae	*Dichroa philippinensis* Schltr.	A. D. E. Elmer 16177 (Luzon, Philippines) NCLC-W 2161
319	Cannabaceae	*Celtis cerasifera* C. K. Schneid.	G. Forrest 24471 (E Tibet/SW China) NCLC-W 9000
320	Salicaceae	*Phylloclinium paradoxum* Baill.	Achten 560 (Luebo, Congo) NCLC-W 7830
322	Betulaceae	*Carpinus laxiflora* (Siebold & Zucc.) Blume	C. Y. Chiao 14466 (Chekiang, China) NCLC-H 6212
323	Chloranthaceae	*Chloranthus serratus* Roem. & Schult.	P. H. Dorsett & W. J. Morse 503 (Fujiyama, Japan) NCLC-H 658
324	Martyniaceae	*Martynia annua* L. ex Rehm.	I. S. Brandegee s.n. (Sinaloa, Mexico) NCLC-H 1706
325	Onagraceae	*Lopezia lopezoides* (Hook. & Arn.) Plitmann, P. H. Raven & Breedlove	McVaugh 14350 (Jalisco, Mexico) NCLC-H 1909)
326	Onagraceae	*Fuchsia decidua* Standl.	D. E. Breedlove 15821 (Guerrero, Mexico) NCLC-H 3852
327	Sapindaceae	*Acer negundo* L.	C. L. Porter 3887 (Colorado, USA) NCLC-W 14573
328	Sapindaceae	*Cupania vernalis* Cambess.	S. Venturi 5206 (Jujuy, Argentina) NCLC-H 2091
329	Celastraceae	*Elaeodendron glaucum* Pers.	S. Ripley 67 (Sri Lanka) NCLC-H 4425

Fig.	Family	Genus and species	Collector and field number (where collected) slide no.
330	Fagaceae	*Quercus alba* × *velutina*	without collector NCLC-W 1079
331	Violaceae	*Melicytus Fasciger* Gillespie	Stauffer 5827 (Fiji) NCLC-W 3246
332	Platanaceae	*Platanus oaxacana* Standley	E.W. Nelson 540 (Mexico) NCLC-H 3743
333	Berberidaceae	*Diphylleia grayi* F. Schmidt	(without collector) (Shinano, Japan) NCLC-H 1168B
334	Vitaceae	*Vitis cavaleriei* H. Lév & Vaniot	(without collector) Maire 7462 (Yunnan, China) NCLC-W 289
335	Iteaceae	*Itea macrophylla* Wall.	Lei 541 (Hainan, China) NCLC-H 3250
336	Malvaceae	*Melochia lupulina* Sw.	D. R. Harris 11955 (Virgin Islands, USA) NCLC-H 5555
337	Onagraceae	*Circaea erubescens* Franch. & Sav.	P. Raven s.n. (Japan) NCLC-H 2154
338	Salicaceae	*Homalium racemosum* Jacq.	E. L. Ekman 7984 (Pinar del Río, Cuba) NCLC-H 1019
339	Trochodendraceae	*Tetracentron sinense* Oliv.	(without collector) (China) NCLC-H 184
340	Aquifoliaceae	*Ilex dipyrena* Wall.	G. Forrest 20680 (Yunnan, China) NCLC-H 4342
341	Salicaceae	*Trimeria alnifolia* Harv.	Rogers 18117 (Transvaal, South Africa) NCLC-H 1016
342	Theaceae	*Hartia sinensis* Dunn.	(without collector) (England, cult.) NCLC-H 5
343	Dilleniaceae	*Schumacheria castaneifolia* Vahl	S. Sohmer & Waas (Sri Lanka) NCLC-H 6793
344	Salicaceae	*Idesia polycarpa* Maxim.	(without collector) (Japan) NCLC-H 1005
345	Cercidiphyllaceae	*Cercidiphyllum genetrix* (Newberry) Hickey	L. Hickey (Golden Valley Fm.) USNM 43234

Fig.	Family	Genus and species	Collector and field number (where collected) slide no.
APP 1	Malvaceae	*Tilia mandshurica* Rupr.	C. Y. Chiao 2721 (Shantung, China) NCLC-H 5406
APP2	Dilleniaceae	*Davilla rugosa* Poir.	E. G. Holt & Gehriger s.n. (Amazonas, Venezuela) NCLC-H 845
APP 3	Dipterocarpaceae	*Stemonoporus nitidus* Thw.	P. S. Ashton 2003 (Sri Lanka) NCLC-H 4665
APP 4	Fabaceae-Caesalpinioideae	*Bauhinia madagascariensis* Desv.	Brion 1843 (Madagascar) NCLC-W 5733
APP 5	Trochodendraceae	*Tetracentron sinense* Oliv.	W. P. Fang 6705 (Szechuan,China) NCLC-W 6550
APP 6	Anacardiaceae	*Buchanania arborescens* (Blume) Blume	Reynoso et al. s.n. (PPI 1403) (Philippines) NY
APP 7	Elaeocarpaceae	*Aristotelia racemosa* Hook.f.	L. Hickey s.n. (New Zealand) NCLC-H 6479
APP 8	Malvaceae	*Bombacopsis rupicola* Robyns	L. Williams 11630 (Venezuela) NCLC-H 5493
APP 9	Gesneriaceae	*Rhynchoglossum azureum* (Schltdl.) B. L. Burtt.	D. E. Breedlove 1154 (Chiapas, Mexico) NCLC-H 1714
APP 10	Nothofagaceae	*Nothofagus procera* Oerst.	P. Moreau 62822 (Argentina) NCLC-H 1760
APP 11	Sapindaceae	*Acer franchetii* Pax	Fang 3924 (Szechuan) NCLC-W 7628
APP 12	Malpighiaceae	*Tetrapterys macrocarpa* I. M. Johnst.	C. O. Erlanson 405 (Panama) NCLC-H 2479
APP 13	Cunoniaceae	*Eucryphia glutinosa* (Poepp. & Endl.) Baill.	Aravena (Linares, Chile) NCLC-W 2468
APP 14	Chrysobalanaceae	*Licania michauxii* Prance	Biltmore, 19496 (Florida, USA) NCLC-H 4026
APP 15	Moraceae	*Morus microphylla* Buckley	Wiggins 7033 (Sonora, Mexico) NCLC-W 14883B
APP 16	Anacardiaceae	*Comocladia dodonaea* (L.) Urban	T. Zanoni et al. 30780 (Dominican Republic) NY

Fig.	Family	Genus and species	Collector and field number (where collected) slide no.
APP 17	Anacardiaceae	*Sorindeia gilletii* De Wild	J. M. & B. Reitsma 3112 (Gabon) NY
APP 18	Proteales	*Leepierceia preartocarpoides* (Brown) Johnson	Johnson 571 (Hell Creek Fm.) DMNH 6359

Appendix D. Instructions for Clearing Leaves

Leaf clearing is the process of removing all pigment and then staining a leaf so that its vein architecture is clearly visible. This procedure can be used on leaves removed (with permission) from herbarium sheets or on live material. Many methods are used for clearing leaves; here we briefly describe one method. The following sources contain additional information on leaf clearing techniques: Foster (1953), de Strittmatter (1973), Hickey (1973), Shobe and Lersten (1967), Pane (1969), and Bohn et al. (2002). Note that this process must be performed in a well-ventilated area because some of the chemicals are harmful to humans.

Leaves are placed in glass containers, covered by a piece of fiberglass mesh to facilitate changing solutions, and submerged in 1–5% NaOH, the strength depending on the thickness of the material. The NaOH solution is changed every 1–2 days during the clearing process, which generally takes 2–10 days. The clearing process is finished by a wash in commercial Clorox® (typically 5–30 seconds) followed by a final wash in water to stop the bleaching process. Clorox removes any remaining pigment from the leaves in preparation for staining. This step requires caution because the leaves are typically fragile from the NaOH treatment and may disintegrate if bleached for too long.

Acid fuchsin is a particularly successful stain, although safranin dye can also be useful. Staining with acid fuchsin involves washing the leaves in 50% ethanol, staining them in 1% acid fuchsin for 3–8 minutes, and then putting the leaves through a dehydration series in 50%, 95%, and 100% ethanol. The first two dehydration steps destain the leaves because the water-soluble dye diffuses out of the leaf into the ethanol; the third step stops the process once there is proper contrast between leaf lamina and stained veins. The specimens can then be rinsed in clove oil, then xylene (a toxic solvent), and finally stored temporarily in a solution of 1:1 xylene:HemoDe®. Proceeding directly from dehydration to storage in HemoDe® also gives good results, but the leaves will eventually lose some pigment.

For photography, the leaves are floated in a glass dish placed on the backlit platform of a dissecting microscope with a digital camera attachment. The acid fuchsin dye fades over time, and limited restaining may be necessary in order to attain the necessary contrast for imaging. Leaves are then permanently mounted on glass slides using standard anatomical techniques. The leaves used in this publication were permanently mounted and then photographed using a light table or converted enlarger condenser as a source of transillumination.

References

Ash, A. W., B. Ellis, L. J. Hickey, K. R. Johnson, P. Wilf, and S. L. Wing. 1999. *Manual of leaf architecture: Morphological description and categorization of dicotyledonous and net-veined monocotyledonous angiosperms.* Washington, D.C.: Smithsonian Institution (http://www.peabody.yale.edu/collections/pb/MLA).

Bailey, I. W., and E. W. Sinnott. 1915. A botanical index of Cretaceous and Tertiary climates. *Science* 46:831–834.

———. 1916. The climatic distribution of certain types of angiosperm leaves. *American Journal of Botany* 3:24–39.

Barnes, R. W., R. S. Hill, and J. C. Bradford. 2001. The history of Cunoniaceae in Australia from macrofossil evidence. *Australian Journal of Botany* 49:301–320.

Basinger, J. F., and D. L. Dilcher. 1984. Ancient bisexual flowers. *Science* 224:511–513.

Bell, A. D. 1991 (reprinted 1998). *Plant form—an illustrated guide to flowering plant morphology.* Oxford: Oxford University Press.

Bohn, S., B. Andreotti, S. Douady, J. Munzinger, and Y. Couder. 2002. Constitutive property of the local organization of leaf venation networks. *Physical Review E* 65:1–12.

Boucher, L. D., S. R. Manchester, and W. S. Judd. 2003. An extinct genus of Salicaceae based on twigs with attached flowers, fruits, and foliage from the Eocene Green River Formation of Utah and Colorado, USA. *American Journal of Botany* 90:1389–1399.

Burnham, R. J. 1994. *Paleoecological and floristic heterogeneity in the plant-fossil record—an analysis based on the Eocene of Washington.* U.S. Geological Survey Bulletin 2085B:1–25.

Cain, S. A., and G. M. D. O. Castro. 1959. *Manual of vegetation analysis.* New York: Harper and Row.

Candela, H., A. Martinez-Laborda, and J. L. Micol. 1999. Venation pattern formation in *Arabidopsis thaliana* leaves. *Developmental Biology* 205:205–216.

Canny, M. J. 1990. What becomes of the transpiration stream? *New Phytologist* 114:341–368.

Carpenter, R. J., R. S. Hill, D. R. Greenwood, A. D. Partridge, and M. A. Banks. 2004. No snow in the mountains: Early Eocene plant fossils from Hotham Heights, Victoria, Australia. *Australian Journal of Botany* 52:685–718.

Chaney, R. W., and E. I. Sanborn. 1933. *The Goshen flora of west central Oregon.* Carnegie Institution of Washington publication 439.

Conran, J. G., and D. C. Christophel. 1999. A redescription of the Australian Eocene fossil monocotyledon *Petermanniopsis* (Lilianae: aff. Petermanniaceae). *Transactions of the Royal Society of South Australia* 123:61–67.

Couder, Y., L. Pauchard, C. Allain, M. Adda-Bedia, and S. Douady. 2002. The leaf venation as formed in a tensorial field. *European Physical Journal B* 28:135–138.

Crane, P. R., and R. Stockey. 1985. Growth and reproductive biology of *Joffrea speirsii* gen. et sp. nov., a *Cercidiphyllum*-like plant from the Late Paleocene of Alberta, Canada. *Canadian Journal of Botany* 63:340–364.

Crepet, W. L., D. C. Nixon, and M. A. Gandolfo. 2004. Fossil evidence and phylogeny: The age of major angiosperm clades based on mesofossil and macrofossil evidence from Cretaceous deposits. *American Journal of Botany* 91:1666–1682.

Cronquist, A. C. 1981. *An integrated system of classification of flowering plants.* New York: Columbia University Press.

Davis, C. C., C. O. Webb, K. J. Wurdack, C. A. Jaramillo, and M. J. Donoghue. 2005. Explosive radiation of Malpighiales supports a mid-Cretaceous origin of modern tropical rainforests. *American Naturalist* 165:E36–E65.

DeVore, M. L., S. M. Moorer, K. B. Pigg, and W. C. Wehr. 2004. Fossil *Neviusia* leaves (Rosaceae: Kerrieae) from the lower-middle Eocene of southern British Columbia. *Rhodora* 106:197–209.

Dickinson, T. A., W. H. Parker, and R. E. Strauss. 1987. Another approach to leaf shape comparisons. *Taxon* 36:1–20.

Dilcher, D. L. 1963. Cuticular analysis of Eocene leaves of *Ocotea obtusifolia. American Journal of Botany* 50:1–8.

———. 1973. A revision of the Eocene flora of southeastern North America. *Paleobotanist* 20:7–18.

———. 1974. Approaches to the identification of angiosperm leaves. *Botanical Review* 40:1–158.

Dilcher, D. L., and P. R. Crane. 1984. *Archaeanthus*: An early angiosperm from the Cenomanian of the western interior of North America. *Annals of the Missouri Botanical Garden* 71:351–383.

Dimitriov, P., and S. W. Zucker. 2006. A constant production hypothesis guides leaf venation patterning. *Proceedings of the National Academy of Science* 103:9363–9368.

Dizeo de Strittmatter, C. G. 1973. Una nueva técnica de diafanización. *Boletín de la Sociedad* Argentina Botánica 15:126–129.

Doyle, J. A. 2007. Systematic value and evolution of leaf architecture across the angiosperms in light of molecular phylogenetic analyses. *Courier Forschungs-Institut Senckenberg* 258:21–37.

Espinosa, D., J. Llorente, and J. J. Morrone. 2006. Historical biogeographical patterns of the species of *Bursera* (Burseraceae) and their taxonomic implications. *Journal of Biogeography* 33:1945–1958.

Feild, T. S., T. L. Sage, C. Czerniak, and W. J. D. Iles. 2005. Hydathodal leaf teeth of *Chloranthus japonicus* (Chloranthaceae) prevent guttation-induced flooding of the mesophyll. *Plant Cell and Environment* 28:1179–1190.

Friis, E. M., K. R. Pedersen, and P. R. Crane, 2006. Cretaceous angiosperm flowers: Innovation and evolution in plant reproduction. *Palaeogeography, Palaeoclimatology, Palaeoecology* 232:251–293.

Friis, E. M., and A. Skarby. 1982. *Scandianthus* gen. nov., angiosperm flowers of saxifragalean affinity from the Upper Cretaceous of southern Sweden. *Annals of Botany* 50:569–583.

Foster, A. S. 1953. Techniques for the study of venation patterns in the leaves of angiosperms. *Proceedings of the International Botanical Congress* 1950:586–587.

Foster, A. S. 1956. Plant idioblasts; remarkable examples of cell specialization. *Protoplasma* 46:184–193.

Fuller, D. Q., and L. J. Hickey. 2005. Systematics and leaf architecture of the Gunneraceae. *Botanical Review* 71:295–353.

Gentry, A. H. 1993. *A field guide to the families and genera of woody plants of northwest South America (Colombia, Ecuador, Peru), with supplementary notes on herbaceous taxa.* Washington, D.C.: Conservation International.

González, C. C., M. A. Gandolfo, and N. R. Cúneo. 2004. Leaf architecture and epidermal characteristics of the Argentinean species of Proteaceae. *International Journal of Plant Sciences* 165:521–526.

Gutiérrez, D. G., and L. Katinas. 2006. To which genus of Asteraceae does *Liabum oblanceolatum* belong? Vegetative characters have the answer. *Botanical Journal of the Linnean Society* 150:479–486.

Hay, A., and M. Tsiantis. 2006. The genetic basis for differences in leaf form between *Arabidopsis thaliana* and its wild relative *Cardamine hirsuta*. *Nature Genetics* 38:942–947.

Herendeen, P. S., S. Magallón-Puebla, R. Lupia, P. R. Crane, and J. Kobylinska. 1999. A preliminary conspectus of the Allon flora from the Late Cretaceous (Late Santonian) of central Georgia, USA. *Annals of the Missouri Botanical Garden* 86:407–471.

Hewson, H. J. 1988. *Plant indumentum: A handbook of terminology.* Canberra: Australian Government Publishing Service.

Hickey, L. J. 1973. Classification of the architecture of dicotyledonous leaves. *American Journal of Botany* 60:17–33.

———. 1974. Clasificación de la arquitectura de las hojas de dicotyledoneas. *Boletín de la Sociedad Argentina de Botánica* 16:1–26.

———. 1977. *Stratigraphy and paleobotany of the Golden Valley Formation (Early Tertiary) of western North Dakota.* Geological Society of America Memoir 150.

———. 1979. A revised classification of the architecture of dicotyledonous leaves. In *Anatomy of the dicotyledons,* 2d ed., ed. C. R. Metcalfe and L. Chalk, vol. 1, pp. 25–39. Oxford: Clarendon Press.

Hickey, L. J., and R. K. Peterson. 1978. *Zingiberopsis,* a fossil genus of the ginger family from Late Cretaceous to early Eocene sediments of western interior North America. *Canadian Journal of Botany* 56:1136–1152.

Hickey, L. J., and R. W. Taylor. 1991. The leaf architecture of *Ticodendron* and the application of foliar characters in discerning its relationships. *Annals of the Missouri Botanical Garden* 78:105–130.

Hickey, L. J., and J. A. Wolfe. 1975. The bases of angiosperm phylogeny: Vegetative morphology. *Annals of the Missouri Botanical Garden* 62:538–589.

Hill, R. S. 1982. The Eocene megafossil flora of Nerriga, New South Wales, *Australia. Palaeontographica Abteilung B. Palaeophytologie* 181:44–77.

———. 1988. Australian Tertiary angiosperm and gymnosperm leaf remains—an updated catalogue. *Alcheringa* 12:207–219.

Högermann, C. 1990. Leaf venation. In *Stratification of tropical forests as seen in leaf structure,* ed. B. Rollet, C. Högermann, and I. Roth, part 2, pp. 77–183. Boston: Kluwer Academic.

Jacobs, B. F., and P. S. Herendeen. 2004. Eocene dry climate and woodland vegetation in tropical Africa reconstructed from fossil leaves from northern Tanzania. *Palaeogeography, Palaeoclimatology, Palaeoecology* 213:115–123.

Jacobs, B. F., and C. H. S. Kabuye. 1989. An extinct species of *Pollia* Thunberg (Commenlianaceae) from the Miocene Ngorora formation, Kenya. *Review of Paleobotany and Palynology* 59:67–76.

Jensen, R. J. 1990. Detecting shape variation in oak leaf morphology: A comparison of rotational-fit methods. *American Journal of Botany* 77:1279–1293.

Johnson, K. J., and B. Ellis. 2002. A tropical rainforest in Colorado 1.4 million years after the Cretaceous-Tertiary boundary. *Science* 296:2379–2383.

Keating, R. C., and V. Randrianasolo. 1988. The contribution of leaf architecture and wood anatomy to classification of the Rhizophoraceae and Anisophyllaceae. *Annals of the Missouri Botanical Garden* 75:1343–1368.

Keller, R. 2004. *Identification of tropical woody plants in the absence of flowers—a field guide.* 2d ed. Basel: Birkhäuser.

Kerner von Marilaun, A. J. 1895. *The natural history of plants: Their forms, growth, reproduction, and distribution,* trans. and ed. by F. W. Oliver. New York: H. Holt. Original German ed.: *Pflanzenlaben.* Leipzig: Verlag des Bibliogräphischen Instituts, 1887–1891.

Kvaček, Z., and S. R. Manchester. 1999. *Eostangeria* Barthel (extinct Cycadales) from the Paleogene of western North America and Europe. *International Journal of Plant Sciences* 160:621–629.

Lam, H. J. 1925. *The Sapotaceae, Sarcospermaceae, and Boerlagellaceae of the Dutch East Indes and surrounding countries.* Bulletin du Jardin Botanique de Buitenzorg ser. II, no. 8.

Levin, G. A. 1986. Systematic foliar morphology of Phyllanthoideae-Euphorbiaceae I. Conspectus. *Annals of the Missouri Botanical Garden* 73:29–85.

Little, S. A., S. Kembel, P. Wilf, D. L. Royer, and B. Cariglino. 2007. Phylogenetic signal in leaf traits used for paleoclimate estimates. In *Abstracts of the Geological Society of America Annual Meeting, Denver, Colo.* 39:22.

MacGinitie, H. D. 1953. *Fossil plants of the Florissant Beds, Colorado.* Carnegie Institution of Washington Contributions to Paleontology no. 559.

Manchester, S. R. 1986. Vegetative and reproductive morphology of an extinct plane tree (Platanaceae) from the Eocene of western North America. *Botanical Gazette* 147:200–226.

Manchester, S. R., and L. J. Hickey. 2007. Reproductive and vegetative organs of *Browniea* gen. n. (Nyssaceae) from the Paleocene of North America. *International Journal of Plant Sciences* 168:229–249.

Manchester, S. R., W. S. Judd, and B. Handley. 2006. Foliage and fruits of early poplars (Salicaceae: *Populus*) from the Eocene of Utah, Colorado, and Wyoming. *International Journal of Plant Sciences* 167:897–908.

Manchester, S. R., K. B. Pigg, and P. R. Crane. 2004. *Palaeocarpinus dakotensis* sp. n. (Betulaceae: Coryloideae) and associated staminate catkins, pollen, and leaves from the Paleocene of North Dakota. *International Journal of Plant Sciences* 165:1135–1148.

Manos, P. S., P. Soltis, D. Soltis, S. Manchester, S. Oh, C. Bell, D. Dilcher, and D. Stone. 2007. Phylogeny of extant and fossil Juglandaceae inferred from the integration of molecular and morphological data sets. *Systematic Biology* 56(3):412–430.

Martínez-Millán, M., and S. R. S. Cevallo-Ferriz. 2005. Arquitectura foliar de Anacardiaceae. *Revista Mexicana de Biodiversidad* 76:137–190.

Meade, C., and J. Parnell. 2003. Multivariate analysis of leaf shape patterns in Asian species of the *Uvaria* group (Annonaceae). *Botanical Journal of the Linnean Society* 143:231.

Melville, R. 1937. The accurate definition of leaf shapes by rectangular coordinates. *Annals of Botany* 1:673–679.

———. 1976. The terminology of leaf architecture. *Taxon* 25:549–561.

Merrill, E. K. 1978. Comparison of mature leaf architecture of three types in *Sorbus* L. (Rosaceae). *Botanical Gazette* 139:447–453.

Meyer, H. W. 2003. *The fossils of Florissant.* Washington, D.C.: Smithsonian Books.

Meyer, H. W., and S. R. Manchester. 1997. *The Oligocene Bridge Creek flora of the John Day Formation, Oregon.* University of California Publications in Geological Sciences no. 141.

Mouton, J. A. 1966. Sur la systematique foliaire en paleobotanique. *Bulletin de la Société Botanique de France* 113:492–502.

———. 1967. Architecture de la nervation foliaire. *Congres national des sociétés savantes* 92:165–176.

O'Dowd, D. J., and M. F. Willson. 1991. Associations between mites and leaf domatia. *Trends in Ecology and Evolution* 6:179–182.

Payne, W. W. 1969. A quick method for clearing leaves. *Ward's Bulletin* 8: 4–5.

Pole, M. 1991. A modified terminology for angiosperm leaf architecture. *Journal of the Royal Society of New Zealand* 21:297–312.

Pole, M., and M. K. MacPhail. 1996. Eocene *Nypa* from Regatta Point, Tasmania. *Review of Palaeobotany and Palynology* 92:55–67.

Raunkiaer, C. 1934. *The life forms of plants and statistical plant geography.* Oxford: Clarendon Press.

Ray, T. S. 1992. Landmark eigenshape analysis: Homologous contours: Leaf shape in *Syngonium* (Araceae). *American Journal of Botany* 79:69–76.

Renner, S. 2004. Bayesian analysis of combined chloroplast loci, using miltiple calibrations, supports the recent arrival of Melastomataceae in Africa and Madagascar. *American Journal of Botany* 91:1427–1435.

Richardson, J. E., M. F. Fay, Q. C. B. Cronk, D. Bowman, and M. W. Chase. 2000. A phylogenetic analysis of Rhamnaceae using *rbcl* and *trnl-f* plastid DNA sequences. *American Journal of Botany* 87:1309–1324.

Richardson, J. E., R. T. Pennington, T. D. Pennington, and P. M. Hollingsworth. 2001. Rapid diversification of a species-rich genus of neotropical rain forest trees. *Science* 293:2242–2245.

Roth, I. 1999. *Microscopic venation patterns of leaves and their importance in the distinction of (tropical) species.* Berlin: Brontraeger.

Roth, I., and V. Mosbrugger. 1999. Architecture and function of angiosperm leaf venation systems—computer simulation studies of the interrelationship between structure and water conduction. In *The evolution of plant architecture,* ed. M. H. Kurmann and A. R. Hemsley, pp. 437–446. Kew, U.K.: Royal Botanic Gardens.

Roth, I., V. Mosbrugger, G. Belz, and H. J. Neugebauer. 1995. Hydrodynamic modeling study of angiosperm leaf venation types. *Botanica Acta* 108:121–126.

Royer, D. L., and P. Wilf. 2006. Why do toothed leaves correlate with cold climates? Gas exchange at leaf margins provides new insights into a classic paleotemperature proxy. *International Journal of Plant Sciences* 167:11–18.

Royer, D. L., P. Wilf, D. A. Janesko, E. A. Kowalski, and D. L. Dilcher. 2005. Correlations of climate and plant ecology to leaf size and shape: Potential proxies for the fossil record. *American Journal of Botany* 92:1141–1151.

Sack, L., E. M. Dietrich, C. M. Streeter, D. Sánchez-Gómez, and N. M. Holbrook. 2008. Leaf palmate venation and vascular redundance confer tolerance of hydraulic disruption. *Proceedings of the National Academy of Science* 105:1567–1572.

Sack, L., and K. Frole. 2006. Leaf structural diversity is related to hydraulic capacity in tropical rain forest trees. *Ecology* 87:483–491.

Sajo, M. G., and P. J. Rudall. 2002. Leaf and stem anatomy of Vochysiaceae in relation to subfamilial and suprafamilial systematics. *Botanical Journal of the Linnean Society* 138(3):339–364.

Seetharam, Y. N., and K. Kotresha. 1998. Foliar venation of some species of *Bauhinia* L. and *Hardwickia binata* Roxb. (Caesalpinioideae). *Phytomorphology* 48:51–59.

Shobe, W. R., and N. R. Lersten. 1967. A technique for clearing and staining gymnosperm leaves. *Botanical Gazette* 128:150–152.

Spicer, R. A. 1986. Pectinal veins: A new concept in terminology for the description of dicotyledonous leaf venation patterns. *Botanical Journal of the Linnean Society* 93:379–388.

Stearn, W. T. 1983. *Botanical Latin: history, grammar, syntax, terminology, and vocabulary.* North Pomfret, Vt.: David & Charles.

Takhtajan, A. L. 1980. 1980. *Outline of the classification of flowering plants (Magnoliophyta).* Botanical Review no. 46.

Taylor, D. W., G. J. Brenner, and S. H. Basha. 2008. *Scutifolium jordanicum* gen. et sp. nov. (Cabombaceae), an aquatic fossil plant from the Lower Cretaceous of Jordan, and the relationships of related leaf fossils to living genera. *American Journal of Botany* 95:340–352.

Theobald, W. L., J. L. Krahulik, and R. C. Rollins. 1979. Trichome description and classification. In *Anatomy of the dicotyledons,* 2d ed., ed. C. R. Metcalfe and L. Chalk, vol. 1, pp. 40–53. Oxford: Clarendon Press.

Todzia, C. A., and R. C. Keating. 1991. Leaf architecture of the Chloranthaceae. *Annals of the Missouri Botanical Garden* 78:476–496.

Tucker, S. C. 1964. The terminal idioblasts in magnoliaceous leaves. *American Journal of Botany* 51:1051–1062.

Uhl, D., S. Klotz, C. Traiser, C. Thiel, T. Itescher, E. Kowalski, and D. L. Dilcher. 2007. Cenozoic paleotemperatures and leaf physiognomy—a European perspective. *Palaeogeography, Palaeoclimatology, Palaeoecology* 248:24–31.

Upchurch, G. 1984. Cuticular anatomy of angiosperm leaves from the Lower Cretaceous Potomac group. *American Journal of Botany* 71:192–202.

Utescher, T., V. Mosbrugger, and A. R. Ashraf. 2000. Terrestrial climate evolution in northwest Germany over the last 25 million years. *Palaios* 15:430–449.

Von Ettingshausen, C. 1861. *Die Blatt-Skelete der Dikotyledonen mit besonderer Rücksicht auf die Untersuchung und Bestimmung der fossilen Pflanzenreste.* Vienna: K. K. Hof- und Staatsdruckerei.

Wang, H., and D. L. Dilcher. 2006. Early Cretaceous angiosperm leaves from the Dakota Formation, Braun Ranch locality, Kansas, USA. *Palaeontographica Abteilung B* 273:101–137.

Webb, L. J. 1959. A physiognomic classification of Australian rain forests. *Journal of Ecology* 47:551–570.

Wilde, V., Z. Kvaček, and J. Bogner. 2005. Fossil leaves of the Araceae from the European Eocene and notes on other aroid fossils. *International Journal of Plant Sciences* 166:157–183.

Wilf, P. 1997. When are leaves good thermometers? A new case for leaf margin analysis. *Paleobiology* 23:373–390.

Wilkinson, H. P. 1979. The plant surface (mainly leaf). In *Anatomy of the dicotyledons,* 2d ed., ed. C. R. Mecalfe and L. Chalk, vol. 1, pp. 97–165. Oxford: Clarendon Press.

———. 1983. Leaf anatomy of *Gluta* (L.) Ding Hou (Anacardiaceae). *Botanical Journal of the Linnean Society* 86:375–403.

Wolfe, J. A. 1971. Tertiary climatic fluctuations and methods of analysis of Tertiary floras. *Palaeogeography, Palaeoclimatology, Palaeoecology* 9:27–57.

———. 1995. Paleoclimatic estimates from Tertiary leaf assemblages. *Annual Review of Earth and Planetary Sciences* 23:119–142.

Wolfe, J. A., and W. Wehr. 1987. Middle Eocene dicotyledonous plants from Republic, northeastern Washington. *U.S. Geological Survey Bulletin* 1597:1–25.

Zamaloa, M. C., M. A. Gandolfo, C. C. González, E. J. Romero, N. R. Cúneo, and P. Wilf. 2006. Casuarinaceae from the Eocene of Patagonia, Argentina. *International Journal of Plant Sciences* 167:1279–1289.

Index